ACROSS CANADA
Resources and Regions
SECOND EDITION

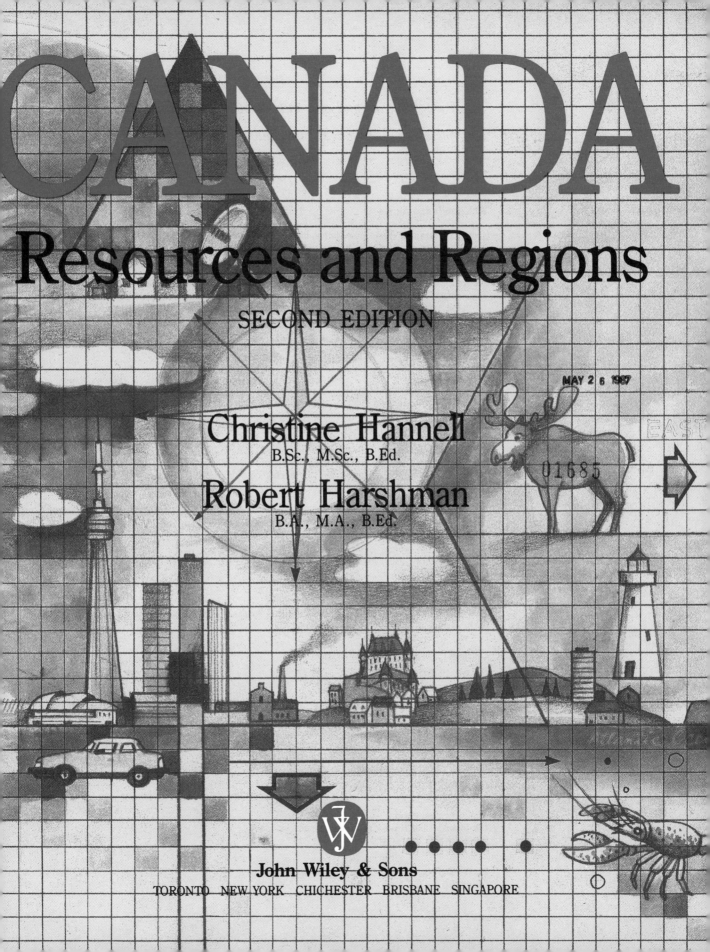

CANADA

Resources and Regions

SECOND EDITION

Christine Hannell
B.Sc., M.Sc., B.Ed.

Robert Harshman
B.A., M.A., B.Ed.

John Wiley & Sons
TORONTO NEW YORK CHICHESTER BRISBANE SINGAPORE

To my wife, Susan, and our daughters, Michelle and Kristen R. Harshman

To Professor F.G. Hannell, our children, and my students C. Hannell

Acknowledgements

Over the years that *Across Canada* has been in use, we have received many suggestions and comments from teachers and students. In this second edition, we have incorporated their valuable suggestions and made changes where appropriate.

We are very grateful to our colleagues and students for their ideas and contributions. In particular, we would like to acknowledge the following three reviewers who have spent many hours going through the text and have given us a thorough evaluation and assessment. Their involvement is much appreciated. They are:

Michael Drouillard, Belle River District High School

Harry Hamill, North Albion Collegiate Institute

Ed Preston, Sir Oliver Mowat Collegiate Institute

Communications Branch, Consumer and Corporate Affairs Canada has granted permission for the use of the National Symbol for Metric Conversion.

Canadian Cataloguing in Publication Data

Hannell, Christine
  Across Canada: resources and regions

For use in Grade 9.
Includes index.
ISBN 0-471-79720-0

1. Canada — Description and travel — 1981 —
I. Harshman, Robert II. Title.

FC75.H35 1987   917.1   C86-094579-0
F1016.H35 1987

Designer: Julian Cleva
Illustrators: Tibor Kovalik and James Loates
Typesetter: Jay Tee Graphics Ltd.

Printed and bound in Canada by The Bryant Press Ltd.
10 9 8 7 6 5 4 3 2 1

CONTENTS

PREFACE

One of the most rewarding things for students is to study about their own country. This enables them to discover more about themselves and their environment as well as the world that awaits them beyond school. In the case of Canada, we have an extraordinarily diverse country. We have people with the national backgrounds of almost every country in the world. In addition, we have natural physical landscapes that vary across the country. It is the interaction of these human and physical elements that makes the geography of the country so fascinating to observers. It is also the focus of this text.

Across Canada is set up so that students can be easily drawn into the study of the geography of Canada. With a multitude of maps, photos, diagrams, and case studies, this book is designed to interest all students in exploring more about Canada.

The text is divided into ten chapters — each one deals with a specific, though at times a general theme. Balance is given to all regions of the country with emphasis on the specific and unique characteristics of each. The thematic approach helps students to understand some of the major forces that help to mould our country as well as hold it together.

One key objective of this text is to familiarize students with the natural resources of Canada and the need to manage them well. Since Canada is a country with vast resources that at one time seemed limitless, it is important that our next generation be aware of the limitations of our environment and the need to manage it wisely. From the forests of Canada to the coastal fisheries to the sprawling cities, Canadians face many new decisions in resource management. The state of our country in the coming century will be determined by the response of Canadians to this issue.

Beyond the general goals of this book, it is hoped that students will find the book comfortable and inviting to use. At the same time students will be challenged in creative ways to understand more about their own country. What they learn about Canada in these lessons will be carried into their later years to influence the progress and shape the future of our country.

Chris Hannell
Rob Harshman

FOREWORD

This second edition of *Across Canada* pleases me even more than the first, which I thought was splendid. Sales of the first edition proved that my judgement was right. Geography texts have become fascinating bits of literature. This new version is even more crammed with detail than its predecessor; but the text still flows as smoothly as ever, and the pages shine even brighter. The book is a worthy successor to the first edition.

What should a geography text do? First and foremost, it should keep the reader awake and excited about the real world around us. More and more the latter is disappearing. The daily experience of the child and later the student is prepackaged and artificial. There is no longer a walk to school, or to college. A bus comes along instead. In a plane, later in life, the student will be told to draw the blinds down, so that the sunlight won't interfere with the movie. And if he or she should want to study a route map, the airline no longer provides one, or offers a very bad substitute that totally misrepresents the world.

Across Canada offsets these backward tendencies. It offers a portrait of a real country, out there for our inspection and comprehension. Canadian national life unfolds on a real stage, made up of soils, rocks, lakes, air, and the living cover. We have worked out a lifestyle that fits what the physical setting demands. Geography is the field that studies that style, that stage, and that reality. It is a concrete field, full of meaning to those who notice the world around them.

To do that demands more than looking idly out of a window. It involves being aware of the ebb and flow of people and things, commodities, and information. Canada's geography throbs with vitality. The authors have sensed that, and have tried to capture some of it within the pages of this book.

F. Kenneth Hare
University of Toronto

1 AN OVERVIEW OF CANADA AND HOW IT RELATES TO THE WORLD

Physical Regions

Canada is a very large and varied country. If you have ever taken a long journey in Canada, you will have noticed some of the differences. You might have seen some beautiful mountains, stark badlands, rich agricultural areas, or wild coastlines. You might have been in an area of large cities, or in a place where your nearest neighbour was many kilometres away. You might have heard different languages spoken and have seen people doing very different jobs.

Using an atlas, locate the place where you live, and describe its general position with respect to Canada as a whole.

Imagine that you are able to take a huge knife and slice Canada across from west to east. You would be able to see the **profile** of the country. A profile is the shape of the land as seen from the side. The profile of southern Canada looks like Figure 1.1.

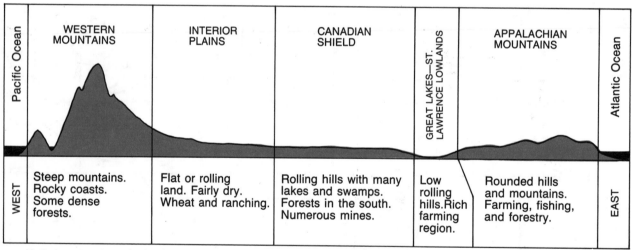

NOTE: Vertical exaggeration varies in different parts of the profile.

Figure 1-1 A profile across southern Canada showing the physical regions

The profile has been divided up into different parts, each called a **region.** A region is an area that has some features that are the same. For example, the Appalachian Mountains is an area of low mountains, uneven coasts, and some productive farmland. The Interior Plains Region has flat or gently rolling land. It has few trees and is used mainly for large ranches or wheat farms.

The second profile (Figure 1.2) shows the shape of the land from the northern Arctic to southern Ontario. As you will notice, the shape of the land is very different from the west-to-east profile.

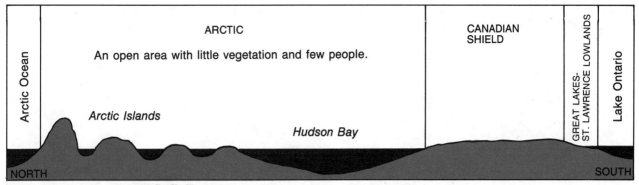

NOTE: Vertical exaggeration varies in different parts of the profile.

Figure 1-2 A profile of Canada from north to south

The map of physical regions, which you can see in Figure 1.3, shows the location of each of the six physical regions. Notice that some regions are much larger than others.

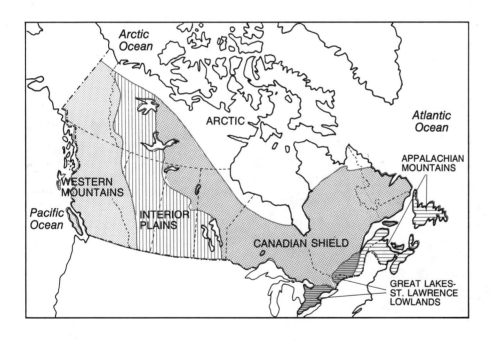

Figure 1-3 A map of the six physical regions of Canada

---

1. (a) In which of the physical regions do you live?
   (b) Briefly describe the location in Canada, and the general landscape of your region. Refer to the profile in Figures 1.1 and 1.2.
2. Photographs showing scenes from each of the six regions of Canada are shown in Figure 1.4. Match each photograph with one region. Give two reasons for each choice.
3. Describe the ways in which the Arctic Region differs from the rest of Canada. Use the profiles, map, and the photographs to help with your answer.

Figure 1-4 Scenes from six physical regions of Canada

Each of Canada's physical regions has a distinctly different appearance, as can be seen from Photographs (a) to (f).

(a)

(b)

(c)

(d)

(e)

(f)

Canada and the World

We started off this chapter by saying that Canada is a very large country. To determine just how large it is we have to find the area of the country, usually expressed as the number of square kilometres it contains. Your atlas will give you this information. We have listed the area of Canada and a few other countries in Figure 1.5. To find others, look in your atlas.

COUNTRY	AREA (km²)	POPULATION (1984)
Argentina	2 777 000	30 100 000
Australia	7 687 000	15 540 000
Canada	9 976 000	25 130 000
China	9 596 000	1 055 304 000
Italy	301 000	57 929 000
Japan	372 000	120 020 000
Netherlands	41 000	14 420 000
Switzerland	41 000	6 440 000
United Kingdom	244 000	56 490 000
U.S.A.	9 363 000	236 680 000
U.S.S.R.	22 402 000	275 000 000

Figure 1-5 The sizes and populations of some countries

4. Which country has a larger area than Canada?
5. Approximately how many times larger is that country?

The table also tells you the **population** (the number of people in each country) in 1984. Most countries count the number of people that they have every ten years, when the year ends in a 1. The proper name given to this counting is a **census**.

6. When will the next census take place?
7. (a) Using information from Figure 1.5, which of the countries listed has a population closest in number to Canada's?
   (b) Although Canada and the other country are similar in population, what difference does the table show?
8. Look at the figures for Canada and the U.S.A. How do these two countries compare in size and in population?

You have probably noticed from looking at these figures that some countries have a large area, but quite a small population (like Canada). Other countries may be small in area, yet have a much larger population (like Japan). The map of the world (Figure 1.6) shows where the people live. Notice how some parts of the world have many people living in them, while other areas have very few people.

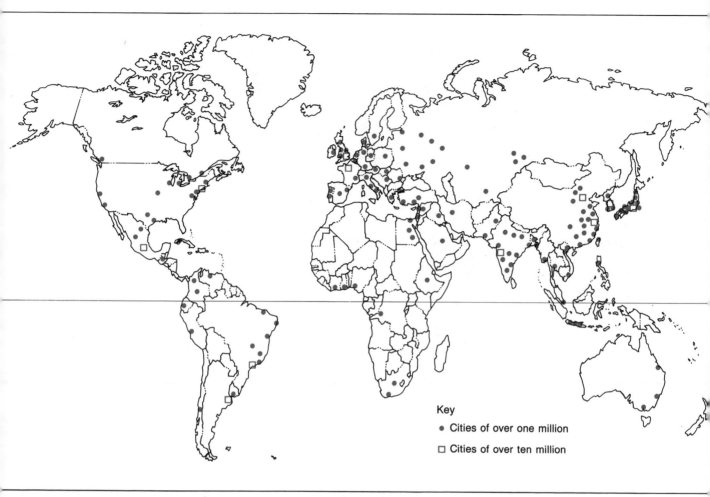

Key

● Cities of over one million

□ Cities of over ten million

Figure 1-6 Cities with more than one million people

In order to measure how many people live in each square kilometre of a country, we work out the **population density.** The population density gives us an idea of the amount of crowding that there is in a country.

Each of the blocks below represents one square kilometre (1 km²) from make-believe countries A, B, C, and D.

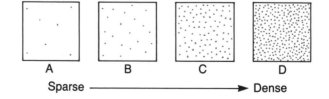

Each dot represents a person. Where there are few people living in each square kilometre, the population is **sparse,** as in A. Where many people are crowded together, as in D, the population is **dense.**

The two photographs (Figures 1.7 and 1.8) show a sparsely and a densely populated area of Canada.

Figure 1-7 An area of dense population *Many Canadian cities have areas like this.*

Figure 1-8 An area of sparse population *Most of Canada is very sparsely settled.*

You can find population density by following this simple method.

Population density (people per square kilometre) = $\dfrac{\text{Total population}}{\text{Total area (km}^2)}$

---

9. Using the formula above, work out the population density of Canada, Japan, the U.S.A., and the Netherlands. Use the data from Figure 1.5 on page 5.
10. In which country would land be most valuable? Explain the reasons for your answer.

One way of comparing statistical information is by using a **bar graph**. In a bar graph, columns, or "bars," are plotted to the right height for each country. Sometimes bar graphs are plotted with the bars going horizontally (left to right).

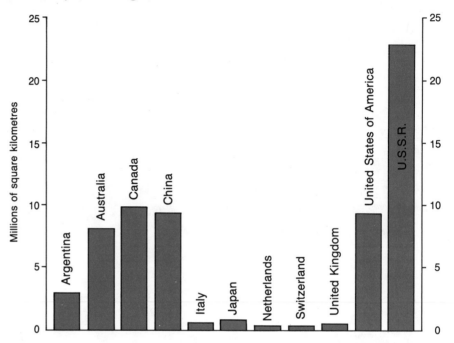

Figure 1-9 A bar graph showing the areas of selected countries

11. Using the figures of population from Figure 1.5 on page 5, plot a bar graph. Use the same method as in the graph shown in Figure 1.9. Use 1 cm for every 50 000 000 people, as you go up the left hand side of the graph. Give your graph a title.
12. Plot a horizontal bar graph using the population density figures that you calculated in answer to Question 9, for Canada, Japan, the U.S.A., and the Netherlands. Give your graph a title.
13. Would you prefer to live in a densely populated part of Canada, or in a sparsely populated area? Give reasons for your answer.

Canada's Trade with Other Countries

Because it is a huge and varied country, Canada has large quantities of valuable minerals, and forest and agricultural products. These are the **raw materials** for many industries. Raw materials are the ingredients needed to make finished products. We use some of the raw materials in Canada, but we also sell a great amount to other countries.

When we sell something to a foreign country, we call it an **export**. We use the money that we get from our exports to buy goods from other countries. The **commodities** (goods) that we buy from other countries are called **imports**.

This buying and selling of goods is called **trade**. When different countries buy and sell, it is called **international trade**.

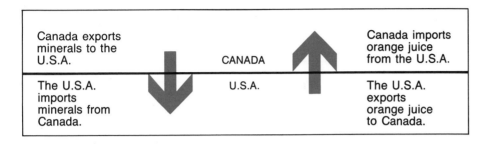

Canada exports minerals to the U.S.A.

CANADA

Canada imports orange juice from the U.S.A.

The U.S.A. imports minerals from Canada.

U.S.A.

The U.S.A. exports orange juice to Canada.

Figure 1-10 International trade

---

14. (a) List five types of food that we import into Canada.
    (b) Why do we have to import them?

The table of figures below (Figure 1.11) shows the value of Canada's exports and imports.

In 1977, for example, we exported goods worth $44 554 000 000, and imported goods worth $42 332 000 000. Because we exported more than we imported, we had a balance of $2 222 000 000 left over. The difference between our exports and our imports is called a **trade balance**. When we sell more than we buy, it is called **a trade surplus**. When we buy more than we sell, it is called a **trade deficit**. If Canada has a trade deficit, it must borrow money to buy the excess goods from other countries. We lose money by doing this. After a few years with a large trade deficit, Canada would have serious economic problems, such as unemployment.

YEAR	EXPORTS	IMPORTS	TRADE BALANCE	DEFICIT OR SURPLUS
1977	44 554	42 332	2 222	Surplus
1978	52 842	49 938		
1979	64 317	62 871		
1980	74 259	69 128		
1981	80 895	78 665		
1982	90 964	75 485		
1983	84 403	67 630		
1984	112 100	91 700		

Figure 1-11 Canada's trade balance (in millions of dollars)

---

15. (a) Copy the table of Canada's trade balance into your notebook.
    (b) Work out the trade balance for each of the years that is left blank.
    (c) Write in whether the balance was a surplus or a deficit for each year.
    (d) In which year did Canada's economy appear to be healthiest? Explain your answer.

9

16. Copy the following graph into your notebook. A *line graph* has been drawn to show how our exports have grown between 1977 and 1984. Use the import figures in Figure 1.11 to draw another line on the graph. Write "imports" on your line.

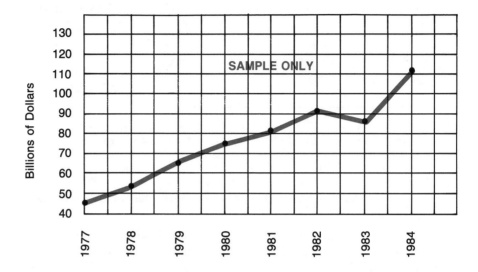

Figure 1-12 Value of exports

17. Describe what has happened to the value of our yearly exports and imports between 1977 and 1984.

Countries usually try to avoid trade deficits. One way that we can avoid a deficit is to cut down on the value of our imports. We could refuse to accept other countries' goods. The problem with this suggestion is that they may say that they will not buy Canadian exports. What is often used is a trade tariff. A trade tariff is a tax put on imported goods. This tax is paid to the Canadian government.

You can see from Figure 1.13 how a tariff increases the cost of an imported product. The tariff may increase the price of the knife so much that you decide to buy a Canadian-made product instead. That is the reason for having tariffs.

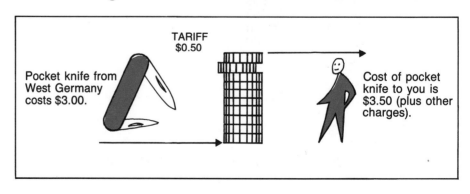

Figure 1-13 A tariff in international trade

Another example is shown in Figure 1.14. The example involves two shirts, identical in style and quality. One shirt is made in Taiwan and sells there for $10. The other shirt is made in Montreal and sells in Canada for $12.

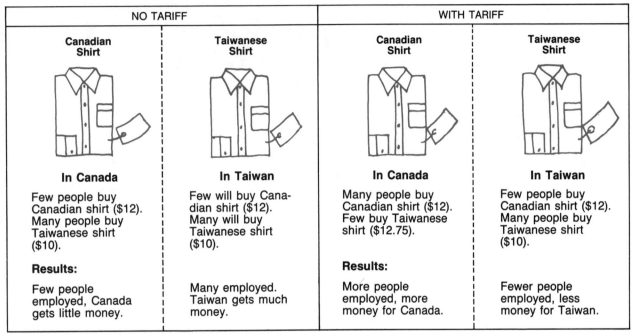

NO TARIFF		WITH TARIFF	
Canadian Shirt	**Taiwanese Shirt**	**Canadian Shirt**	**Taiwanese Shirt**
In Canada	**In Taiwan**	**In Canada**	**In Taiwan**
Few people buy Canadian shirt ($12). Many people buy Taiwanese shirt ($10).	Few will buy Canadian shirt ($12). Many will buy Taiwanese shirt ($10).	Many people buy Canadian shirt ($12). Few buy Taiwanese shirt ($12.75).	Few people buy Canadian shirt ($12). Many people buy Taiwanese shirt ($10).
Results:		**Results:**	
Few people employed, Canada gets little money.	Many employed. Taiwan gets much money.	More people employed, more money for Canada.	Fewer people employed, less money for Taiwan.

NOTE: These are only sample prices.

Figure 1-14 How a tariff works

Free trade means trade without any tariffs.

---

18. What is a trade tariff?
19. Who is a tariff supposed to help?
20. Who is hurt by tariffs?
21. Do you think that we should have tariffs? Explain the reasons for your answer.
22. What would be the
    (a) advantages
    (b) disadvantages of free trade between Canada and the U.S.A.?

The countries with which Canada trades are called our **trading partners.** The world map in Figure 1.15 has our ten most important trading partners marked on it. The numbers indicate their order of importance to Canada. For example, as you would expect, the U.S.A. is our most important trading partner and is therefore numbered 1 on the map.

Why do we trade with these countries in particular? Many of these countries produce **manufactured goods,** which have been made from raw materials.

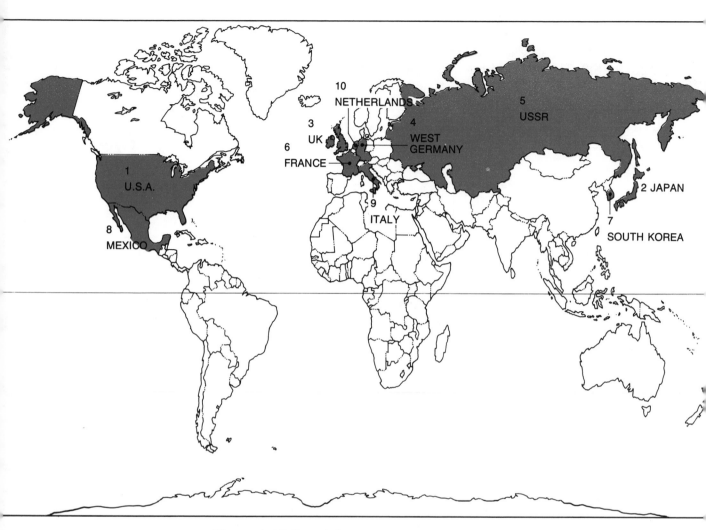

10
NETHERLANDS
3
UK
6
FRANCE
4
WEST
GERMANY
5
USSR
1
U.S.A.
8
MEXICO
9
ITALY
2 JAPAN
7
SOUTH KOREA

Figure 1-15 Canada's ten most valuable trading partners (1984)

Often, many people are needed to make these manufactured goods. Canada does not have very many people to make such products. Also, because we have few people, we find it difficult to get enough **capital** (money) together to build factories.

The countries that send us large quantities of manufactured goods are the U.S.A., Japan, the United Kingdom, West Germany, Italy, and France. In return, we send them large quantities of raw materials. There are also a number of countries that send us raw materials. For example, we import oil to help heat our homes and run our industries in eastern Canada.

23. **Why do you think our trade with most of the countries of Africa and South America is so small?**

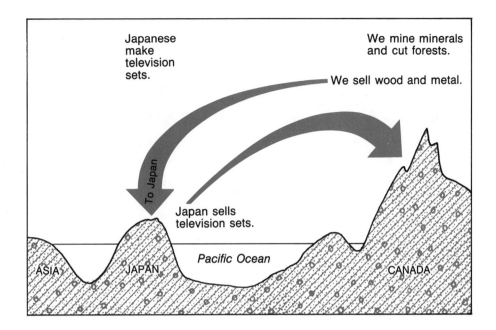

Figure 1-16
International trade
between trading
partners

TRANSPORTATION OF OUR IMPORTS AND EXPORTS

Trading is vital to Canadians if we wish to lead a comfortable life. Canada is in a good location for trading with other countries. Look at a world map in your atlas. You will see that Canada is surrounded by water on three sides. The Pacific Ocean is to the west, the Atlantic Ocean to the east, and the Arctic Ocean to the north. To the south is the U.S.A.

---

24. Using the map of our trading partners (Figure 1.15 on page 12), and a world map in your atlas, list the oceans and seas over which ships would travel between Canada and each of the countries shown.
25. Give two reasons why the Arctic Ocean is not used much as a trade route.
26. Why is it an advantage to be next to the U.S.A. for trading purposes?

Shipping is usually less expensive than rail or road transportation, so when it is possible, ships are used. For example, if you look at the world map on page 12, you will see that we trade a great deal with Mexico. The oil that we import from Mexico could be sent part way by a land route, but it is much cheaper to send it by sea, using tankers.

Trade with the U.S.A. is done mostly overland, although bulky commodities are moved by ship where possible. Rail is the most economical way of moving materials over a land route, but long distance trucks are used for a great amount of our land transportation.

Figure 1-17 Unloading cargo from a ship *Most of Canada's trade with the countries of Asia, Europe, South America, and Africa is by ship.*

Figure 1-18 Ship loading grain *Many ships from Asia load grain in Vancouver Harbour.*

Figure 1-19 Train hauling freight *This train is transporting iron ore from the mines in the Canadian Shield to the steel mills in southern Ontario. It is called a "unit train" because it carries only one type of product.*

Figure 1-20 Long distance truck *Long distance trucks such as this one haul most of the freight in Canada. They can deliver to any place served by a road.*

27. (a) Why do you think that trucks are used more frequently than trains to transport goods in Canada?
    (b) What negative effects result from heavy truck traffic on our highways?

Imagine a large shipment of sewing machines being sent to a small town in southern Saskatchewan from the United Kingdom. It would probably travel across the Atlantic by ship, then continue to Regina by rail. It would then be unloaded and put on a truck to get it to its final destination.

Transhipment is the loading and unloading of commodities from one form of transportation to another. The more quickly and efficiently transhipment can be achieved, the lower the cost. For example, instead of shipping each sewing machine separately, hundreds of them could be packed inside a huge container. This container is then moved from ship to rail, or rail to road, without being unpacked. This saves time and money. Montreal, Halifax, and Vancouver are important ports for such **containerized freight.**

Figure 1-21 Moving a shipment of sewing machines from the United Kingdom to Saskatchewan

Figure 1-22 Containerized freight being unloaded. . . (a) From a ship

(b) **Onto a train**

(c) **Onto a truck**

Very little mention has been made so far of **air cargo** (transporting goods by air). Air cargo is usually the fastest way of transporting objects. It is fairly expensive in comparison to the other methods of transportation. For example, air cargo is twice as expensive as sending the same item by sea. However, it is often used for certain types of commodities. Something which is fairly small, but quite valuable, might be flown into or out of Canada. For example, watches from Switzerland or diamonds from the Netherlands would probably be flown into Canada.

THE IMPORTANCE OF IMPROVEMENTS IN TRANSPORTATION

Improvements in transportation have effectively made the earth seem much smaller. By using available data on the time taken by ship or airplane to cross the Atlantic, Figure 1.23 was constructed.

The consequences of the increased speed of transportation have been great. International trade supplies the needs of industry, and provides a way of getting finished products to the buyers. Our diets have changed. Fast transportation and communication help to bind international organizations together and permit quick military action or relief following disasters. With the increase in speed and comfort, tourism has developed as an important activity of great economic significance.

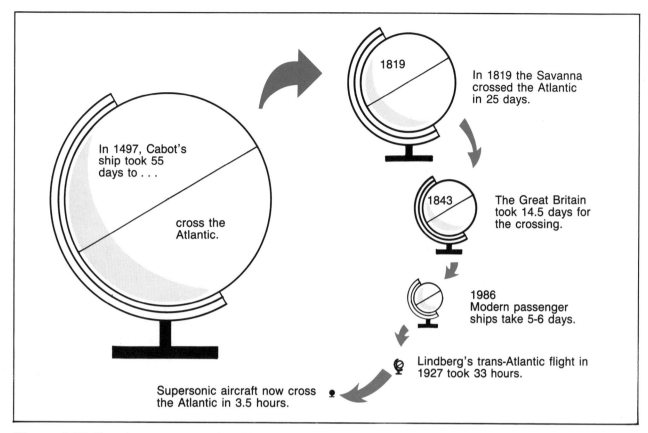

Figure 1-23 How improvements in shipping and air transport have "shrunk" the world in terms of travel time

Latitude and Longitude

The ability to navigate is important to those who have to move goods around the world. There are many different ways of doing this. To be able to guide a ship across wide oceans is different from trying to find your way across the land. On land, features like mountains and rivers may help you to find your way. In the middle of an ocean, there may be no land-marks for hundreds of kilometres. The early explorers worked out a system for finding their location. They took measurements of angles to the sun and stars. Knowing the time of day or night, they were able to calculate their exact location.

The system is very simple. It is like a special net that has been put over the whole world. Each strand of the net has a number. The strands that go around the earth from west to east, parallel to the Equator, are called **lines of latitude.** (Some people call them parallels.)

Latitude is a distance measured in degrees north or south of the Equator. The highest number for latitude is 90°. At 90°N you would find the North Pole. What would you find at 90°S?

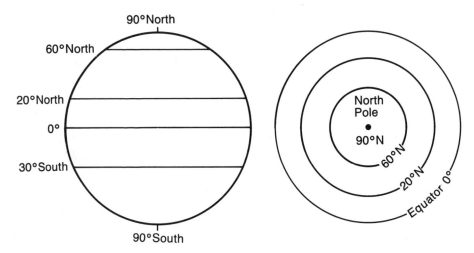

Figure 1-24 Some lines of latitude

Figure 1-25 Lines of latitude seen from above the North Pole

Looking at a globe from above the North Pole you would see the lines of latitude as circles. A few of the lines of latitude are so important that they have a name as well as a number, as you can see in Figure 1.26.

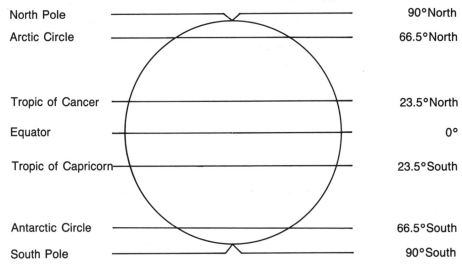

Figure 1-26 Important lines of latitude

Lines of longitude are the strands of the net that pass through the North and South Poles. (Some people call them meridians.)

Longitude is a distance measured west or east from the **prime meridian.** The prime meridian is the 0° line of longitude that passes through Greenwich, England.

Looking at a globe from above the North Pole, you will see the lines of longitude going out like the spokes of a bicycle wheel. The highest number is 180°. (The International Date Line goes along 180° for much of its course.)

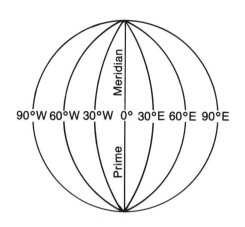

Figure 1-27 Lines of longitude

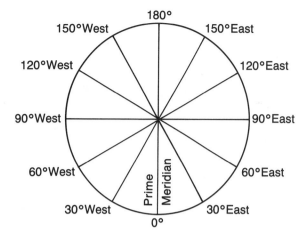

Figure 1-28 Lines of longitude viewed from above the North Pole

In Figure 1.29 you will find a map of the Atlantic Ocean. On it a ship's route is marked.

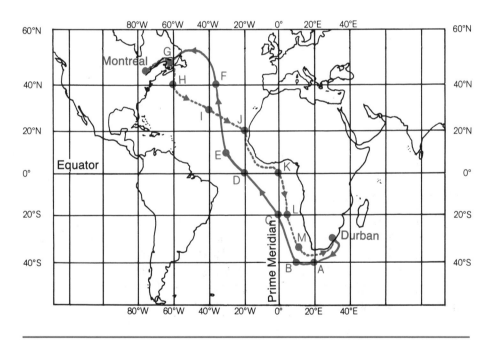

Figure 1-29 A ship's route between Montreal and Durban

28. The navigator took latitude and longitude readings at the places marked A to G. These were his observations.

A 40°S 20°E	D 0° 20°W	G 49°N 62°W
B 40°S 10°E	E 10°N 30°W	
C 20°S 0°	F 40°N 35°W	

On the return journey to Durban, he took a slightly different route and took latitude and longitude readings at H to M. Write down the latitude and longitude readings of each of these places. Remember to write the latitude number first and to read the numbers in the right direction.

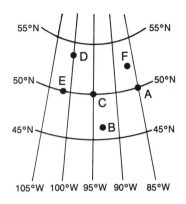

29. (a) Latitude and longitude are also used to locate places on the surface of the land. In your atlas, turn to a map that shows the whole of Canada. Draw the following lines of latitude and longitude in your notebook:
    Latitude: 45°N, 55°N, 65°N, 75°N. Correctly number each line and label each one "latitude."
    Longitude: 55°W, 75°W, 95°W, 115°W, 135°W. Correctly number each line and label each one "longitude."
    (b) Look at the numbers on the lines of latitude in your atlas. Are they marked in degrees north or in degrees south?
    (c) Look at the numbers on the lines of longitude. Are they marked in degrees west or in degrees east?
    (d) When you are counting degrees of latitude in Canada, do you go from south to north, or from north to south?
    (e) When you are counting degrees of longitude in Canada, do you go from west to east, or from east to west?
    (f) How many degrees are there between each printed line of latitude or longitude?

30. On the diagram in the margin, give the latitude and longitude of the dots at C, D, E, and F. A and B have been done for you.
    (a) 50°N 85°W   (b) 47°N 93°W

31. On an atlas map of Canada, find the big city that is close to the following:
    (a) 50°N 97°W   (c) 45°N 66°W
    (b) 49°N 123°W   (d) 44°N 79°W

32. Write the latitude and longitude, to the nearest degree, of the following:
    (a) the farthest point east in Canada   (c) the place where
    (b) the farthest point south in Canada   you live

CANADA'S TIME ZONES

The key to an understanding of time zones in Canada is the rotation of the earth on its axis. The **earth's axis** stretches in a straight line from the North to the South Pole. Slowly, the earth spins around its axis. The half of the world that faces the sun experiences day while the other half is in darkness. As the earth spins, a different part of the world enters day and the opposite side of the world enters night.

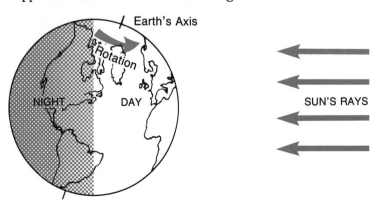

Figure 1-30 Day and night

20

As the earth turns, the east coast of Canada is the first to come into the sun's light. As hours pass, more and more of Canada becomes light. Canada, however, is such a huge country that the sun brings full light to Newfoundland many hours before it is seen in British Columbia and the Yukon. When people on the Pacific coast are starting breakfast, those on the east coast have finished their lunch.

In order to simplify this variation in time across the country, Canada has been divided into six time zones. These time zones are shown in Figure 1.31.

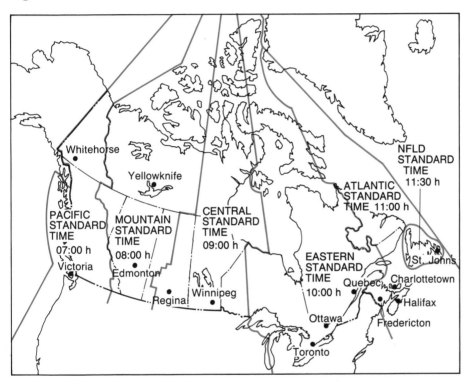

Figure 1-31 Canada's time zones

As you can see from Figure 1.31, time is more advanced in eastern Canada than in western Canada. An understanding of time zones is important to those people who have to travel across Canada, or to communicate with people in other areas.

---

33. Using Figure 1.31 as a guide, answer the following questions:
    (a) Locate the time zone in which you live. What is its name?
    (b) If it is 08:00 h in Halifax, what is the time in Edmonton?
34. (a) If you wanted to watch the Grey Cup, starting at 14:00 h Pacific Time, when should you turn on your television?
    (b) The Santa Claus parade in Toronto starts at 11:00 h. At what time should you start to watch it?
35. Imagine that you are travelling by jet from St. John's to Winnipeg. Your take-off is at 14:00 h and the journey will take three hours. At what time will you land in Winnipeg?

Other Ways in Which Canada Interacts with the World

Now that you have learned something about Canada's international trade, we will finish off this section by briefly mentioning the other ways in which Canada and the world interact. We take part in world affairs through international, political, cultural, and private organizations.

When disaster strikes in part of the world, for example, Canada often helps by sending money, equipment, supplies, and people. Canadian military personnel are also often used to help keep peace in some parts of the world.

Figure 1-32 Nations of the Commonwealth *The Commonwealth consists of thirty-six countries, and includes about one-quarter of the world's population. Well over half of the Commonwealth citizens are very poor. Canada gives aid in many forms to help these people.*

---

36. Collect a city newspaper for four or five days.
    (a) Cut out articles where Canadians in other countries are mentioned.
    (b) State, if possible, which organization is involved in each case.

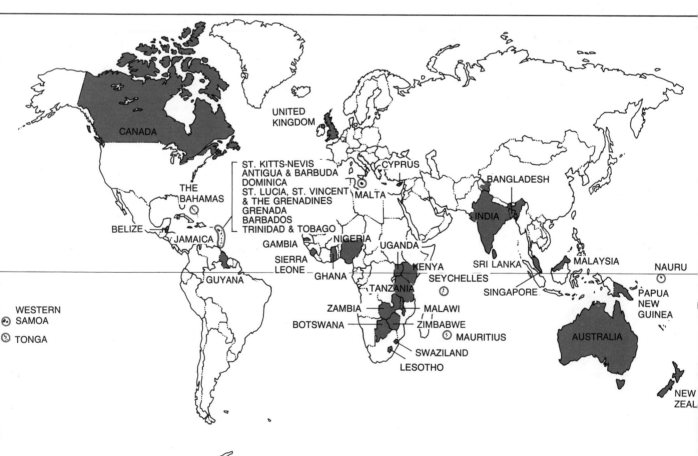

22

The major political organizations to which Canada belongs are the Commonwealth of Nations, the United Nations (UN), the North Atlantic Treaty Organization (NATO), and the North American Defence Pact (NORAD).

The Commonwealth of Nations has developed from the old British Empire. Many member countries, including the United Kingdom, Australia, various African nations, and Canada, have made trading agreements. All participating nations benefit from these agreements. Many people travel and work in the other countries of the Commonwealth.

Figure 1-33 Commonwealth activities *Equipment and advisers from Canada are helping these Zambians to build new roads.*

Most countries in the world belong to the United Nations. The main aims of this organization are to keep peace in the world and to give aid where it is needed.

Figure 1-34 United Nations activities *Canadian peacekeepers patrol the Green Line in Nicosia, Cyprus.*

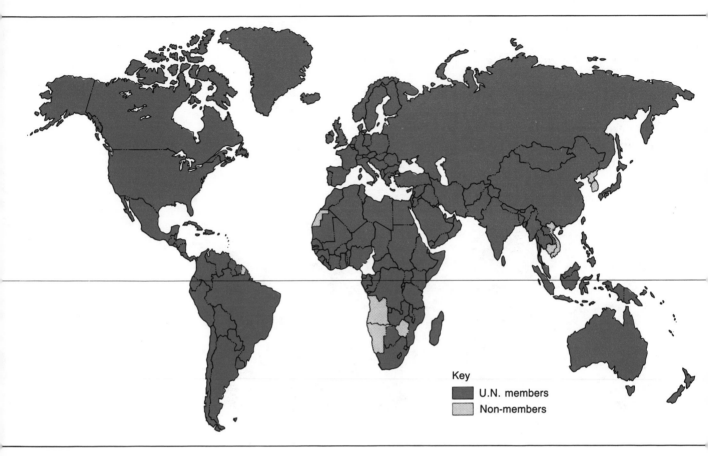

Key
U.N. members
Non-members

Figure 1-35 Countries of the United Nations

The North Atlantic Treaty Organization developed after World War II. It is an organization of many European countries as well as the U.S.A. and Canada. They are mainly concerned about preventing the spread of communism into western Europe.

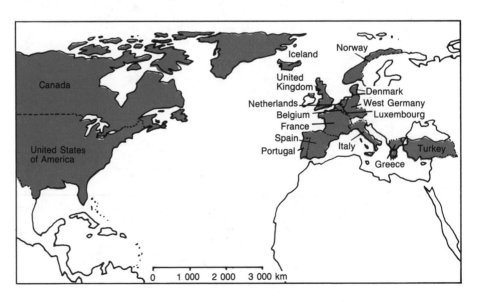

Figure 1-36 Countries of the North Atlantic Treaty Organization

Norad involves only Canada and the U.S.A. If you look at a globe of the world, you will see that Canada is directly between most of the U.S.A. and the U.S.S.R. If a war ever broke out between these two countries, Canada would be in the middle. To give early warning of missiles approaching over the Arctic, radar bases have been built in northern Canada. The organization, therefore, is a form of mutual defence against possible aggression by the U.S.S.R. The line of radar bases is called the DEW Line (Distant Early Warning Line).

Figure 1-37 Shortest route for missiles from the U.S.S.R. to the U.S.A.

Figure 1-38 A DEW-line base

CANADA'S FOREIGN AID

Canada ranks as one of the richest countries in the world. One way to measure this wealth is through the use of the **GNP** (Gross National Product) per person. This means the amount of wealth produced in a country on average per person, during the year. Figure 1.39 lists the *per capita* (per person) GNP for several countries.

COUNTRY	GNP PER PERSON ($ U.S.)	COUNTRY	GNP PER PERSON ($ U.S.)
Australia	11 080	India	260
Brazil	2 220	Nigeria	870
Burundi	230	Spain	5 640
Canada	11 400	Thailand	770
Egypt	650	United States	12 820
Guyana	720	West Germany	13 450

Figure 1-39 The per capita Gross National Product in selected countries (1982)

37. (a) How many times greater than that of Burundi, is Canada's per capita GNP? Divide Burundi's figure into that of Canada to obtain the answer.
    (b) Working from your answer to 37 (a), in what ways would the daily life of a Canadian be different from that of a resident of Burundi?
38. (a) Using your atlas and its index, locate each of the countries listed in Figure 1.39. List the countries according to the continents in which they are found (the world map is given below to help you). Based on this information, which continents have a high GNP per person?
    (b) Which have a low GNP per person?

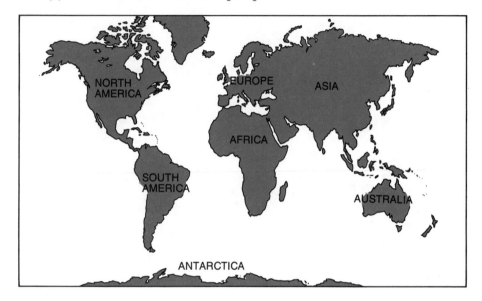

Figure 1-40 Poor housing *Candians help to improve poor housing such as this in Indonesia.*

26

The data from Figure 1.39 indicate how different our way of life is from that of most people of the world. The Canadian government and many individuals in our country have become concerned about the extreme poverty in other parts of the world. This concern has been expressed in the development of foreign aid, or help, for the poor of the world.

Canadian programs for foreign aid are carried out either by the federal government or by private organizations. The government directs its aid through CIDA (Canadian International Development Agency). Other private organizations include OXFAM, CARE, World Vision, and the Red Cross. Such private organizations receive money mainly from individuals rather than from the government.

CIDA is involved in a wide variety of programs throughout the world. Of the $678 300 000 assistance in 1983 to 1984, 44% was allocated to Africa, 41% to Asia, and 14% to the Americas. These programs included the following:
• $50 000 000 to help with emergency food aid in Africa.
• $13 200 000 for small and medium sized industrial projects in Colombia.
• $940 000 to help with foresty, agriculture, energy production, and human resource development in China.

Figure 1-41 The Seventh Pan American Wheelchair Games in Halifax, 1982 *In this game on the final day of play, the U.S. team took the gold medal with a 51-36 win over Canada.*

Figure 1-42 The Expo Centre at Expo 86 in Vancouver *Expo 86 attracted millions of visitors from all over the world.*

Many athletes and tourists visit Canada and many Canadians visit other countries. Important events, such as the Olympic Games in Montreal (1976), the Commonwealth Games in Edmonton (1978), and the Winter Olympics near Calgary (1988) are well known, but there are many smaller international events that take place.

Visitors to Canada are very important to our economy. In 1983, foreign visitors spent $3.8 billion. Together with money spent by Canadian tourists in Canada, the total was nearly $15 billion. Tourism, directly or indirectly, employs a great many people.

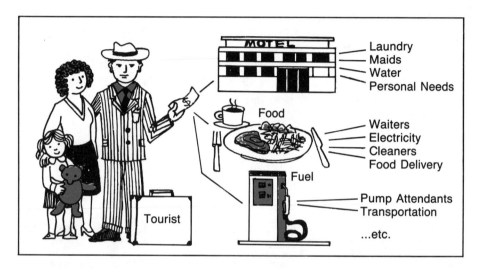

Figure 1-43
Employment generated by tourists *Tourism creates 600 000 jobs directly in Canada, and many more indirectly.*

39. **Draw a table in your notebook like the one below.**

ORGANIZATIONS TO WHICH CANADA BELONGS		
ORGANIZATION	SHORT FORM	MAIN AIMS
Commonwealth of Nations		
United Nations	**SAMPLE ONLY**	
North Atlantic Treaty Organization		
North American Defence Pact		

40. **Fill in the blank part of the table using the information from pages 23-25.**
41. **List some other types of jobs that would be created by tourists.**

Profile
Region
Population
Census
Population density
Sparse
Dense
Bar graph
Raw materials
Export
Commodities

Imports
Trade
International trade
Trade balance
Trade surplus
Trade deficit
Line graph
Trade tariff
Free trade
Trading partners
Manufactured goods

Capital
Transhipment
Containerized freight
Air cargo
Lines of latitude
Latitude
Lines of longitude
Longitude
Prime meridian
GNP

Research Questions

1. On a blank map of Canada, name each of the provinces and territories and their capitals. Mark Ottawa. Make sure that your spelling is correct.
2. List five things that Canada exports. For each one, find out where it is produced in Canada.
3. On a map of Canada, name the most important places that, in your knowledge, attract tourists. For each one, explain what attracts the tourists. Use postcards or photographs, if you have some.
4. Find out what is meant by ''cultural exchange.'' Imagine you were in charge of organizing a cultural exchange with an African nation. What kinds of people would you choose to take with you to represent Canadian traditions?
5. Find out about the organization called ''*la Francophonie.*'' Map the countries which are involved and explain the activities which take place in this organization.

2 CANADA'S PHYSICAL BASE

Landscape

Very few countries of the world can boast of the range of landscape contained within Canada. As well, Canada's climate has a variety as great as that of its landscape. A few basic ideas will help you to understand why we have these differences.

There are two main groups of forces that determine the shape of the surface of the land. These are as follows:
• The forces that create the rock foundation.
• The forces that act on the rocks to change the surface shape.

THE ROCK FOUNDATION

Underneath every part of Canada there is a foundation of hard rocks.

In some places these rocks can be seen very easily, such as in many mountainous areas. In other places you might have to dig down for 20 or 30 m before reaching the rock.

Figure 2-1 Rocks exposed on a mountainside

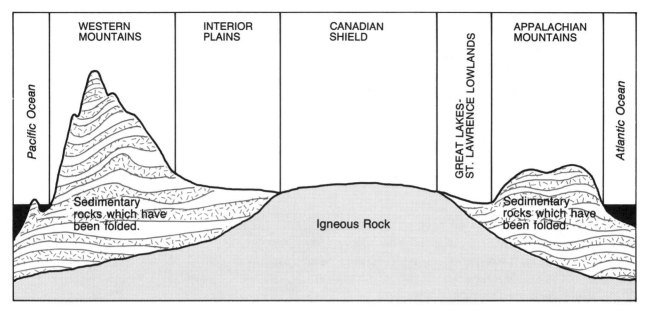

Figure 2-2 Cross section from west to east through southern Canada showing the rock structure *The hard rocks of the Canadian Shield form the "basement" for all of Canada. Areas composed of layered rocks, greatly folded in some places, surround the Shield.*

If you were able to slice through Canada from west to east, the rock pattern you would see would look like the pattern in Figure 2.2.

The Rock Cycle

Before you can fully understand how the rock foundation of Canada was formed, it is necessary to study how rocks are formed. A convenient point to begin this study is when the rocks are in a molten state. The correct name for molten rock is magma. It is believed that the earth was like this when it was first formed. When it cools, this melted rock hardens and turns into igneous rocks, one of which is granite.

Figure 2-3 Granite *This is a rock commonly found in the Canadian Shield.*

Figure 2-4 Solidified magma *The vertical pillars are the remains of tree stumps burned off by the red-hot magma.*

Hills and mountains of igneous rock are weathered and eroded. Weathering is the breaking up and rotting of solid rocks. Erosion is the wearing away of rocks. The small particles that result are moved by rivers to the sea or a lake. They collect in layers on the floor, under the water.

These layers harden and become sedimentary rocks such as sandstone. Sometimes fossils (the hard remains of animals or plants) are preserved in sedimentary rocks.

Figure 2-6 Fossil sea shells in sandstone

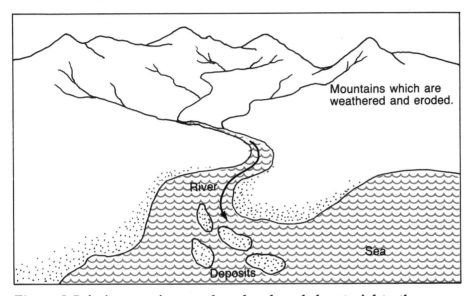

Mountains which are weathered and eroded.

River

Sea

Deposits

Figure 2-5 A river carries weathered and eroded material to the sea

Sometimes the rock is subjected to stresses and strains or heating by volcanoes. This changes igneous or sedimentary rocks into metamorphic rocks, such as gneiss (pronounced nice).

The way in which one rock changes into another is shown in Figure 2.8 below.

Figure 2-7 Gneiss

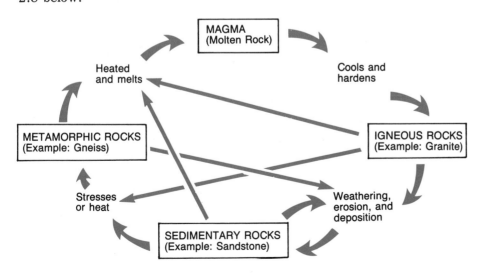

MAGMA
(Molten Rock)

Heated and melts

Cools and hardens

METAMORPHIC ROCKS
(Example: Gneiss)

IGNEOUS ROCKS
(Example: Granite)

Stresses or heat

Weathering, erosion, and deposition

SEDIMENTARY ROCKS
(Example: Sandstone)

Figure 2-8 The rock cycle

1. You will notice from Figure 2.8 that extra lines have been drawn in.
   (a) Which rock type is formed when magma cools?
   (b) Which rocks can melt to form magma?
   (c) Which rocks can be broken down to form sedimentary rocks?
2. Explain, using full sentences, the three changes that could occur to a mass of igneous rock. What would cause each change?

HOW THE ROCK STRUCTURE WAS CREATED

The Canadian Shield

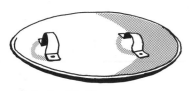

Figure 2-9 The similarity in shape between . . .

(a) A shield

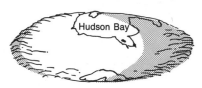

(b) The Canadian Shield

The oldest rocks in Canada are those that make up the Canadian Shield. Some of these rocks are as old as 2 500 000 000 years! They formed from magma when the earth was young. When the rocks hardened, they formed a huge area of igneous rock. The Shield is highest around its edges and lowest near to Hudson Bay. Thus its surface is shaped like a shield.

Great strains forced the rocks of the Canadian Shield into huge mountains. The strains caused some areas of igneous rocks to change into metamorphic rocks. Many of our most valuable mineral deposits, such as iron, silver, and gold, are found in areas of metamorphic rocks.

The Remainder of Canada's Rocks

The mountains of the Shield were slowly worn down. Rivers carried small pieces or rock, sand, and mud to the surrounding seas. These formed layers of sedimentary rocks under the water.

Figure 2-10 The first stages in the formation of Canada's rock foundation

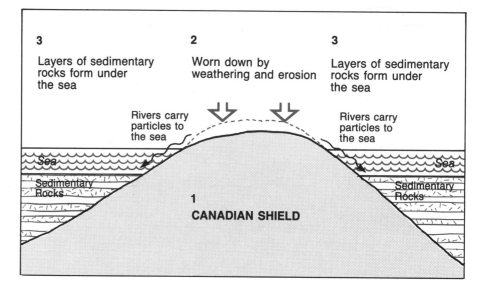

In the east and north, great forces inside the earth folded the sedimentary layers into mountains. The mountains in the east are called the Appalachians. The mountains in the north are called the Innuitians.

Huge forces then pushed up the Western Mountains into folds. These mountains are higher than the Appalachians because they are younger and have not been worn down for as long.

On top of even our highest mountains you would be able to find fossils of animals that were once under the sea. This proves that the rocks that make up these mountains were formed underwater.

The Interior Plains and the Great Lakes–St. Lawrence Lowlands were the last two to be formed. Both regions have sedimentary rocks underneath them. As the seas slowly retreated (drained away), fairly flat layers of sedimentary rocks were left behind at the surface. Figure 2.2 on page 32 shows the present rock structure of Canada.

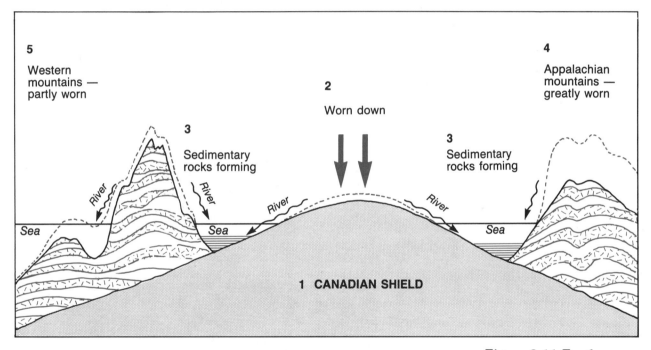

Figure 2-11 Further stages in the development of Canada's rock foundation

3. Arrange these regions in the order in which they were formed, beginning with the oldest:
Western Mountains, Interior Plains, Great Lakes-St. Lawrence Lowlands, Appalachian Mountains, Canadian Shield.

4. (a) Using your atlas, find the greatest heights of the Western Mountains, the Canadian Shield, and the Appalachian Mountains.

(b) Explain the reasons for the differences in the height of the mountains listed in 4(a). Base your answer on the information given to you in the section entitled "How Canada's Rock Structure Was Created."

WEATHERING AND EROSION

Once the basic rock structure has been created, it may be altered, as you can see in the following photographs. Rivers sometimes carry a great quantity of mud. This results from the weathering and erosion of the land. The mud carried by the rivers is eventually deposited on the floor of the sea or ocean.

The rock in Figure 2.13 has been severely weathered. This weathering results from the heating action of the sun, or the freezing and thawing of water in the cracks of the rock.

Once rock chips have been produced by weathering, they may move down a slope such as the one in Figure 2.14. They may collect in a pile at the base. Loose particles may also be washed away by rain or, if they are dry and light enough, they may be blown by the wind.

Figure 2-12 A river carrying mud to the sea

Figure 2-14 Rock pieces piling up at the base of a cliff

Figure 2-13 Weathered rock

Figure 2-15 Wind erosion exposes roots of trees

About 10 000 years ago, thick masses of ice called **glaciers** moved across the surface of Canada. In many parts of Canada, such as in the Western Mountains, the glaciers carved spectacular scenery.

Figure 2-16 A glacier

Figure 2-17 Glaciated scenery in the Western Mountains

Figure 2-18 Lakes in the Canadian Shield

Figure 2-19 Rich farmland of the Great Lakes– St. Lawrence Lowlands

Figure 2-20 Destructive effects of waves

In the Canadian Shield, glaciers carved out many thousands of hollows in which lakes have now formed. The glaciers often moved millions of tonnes of loose particles. When the ice melted, these particles were left behind. They became the foundation for deep soils, such as are found in the Great Lakes–St. Lawrence Lowlands.

Along our coasts, waves perform two basic functions. In some cases, they wear away the cliffs. In other places, they build up beaches.

The shape of the surface of the land depends on the balance between the forces that build up the land and the forces that lower the land. Changes in our landscape usually occur very slowly and continuously.

Figure 2-21 The major processes of landscape formation

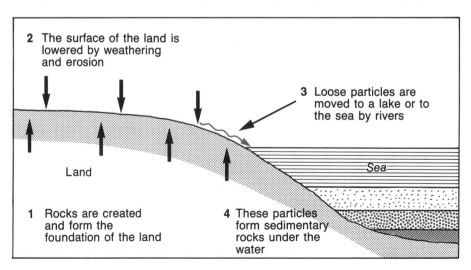

2 The surface of the land is lowered by weathering and erosion

3 Loose particles are moved to a lake or to the sea by rivers

Land

Sea

1 Rocks are created and form the foundation of the land

4 These particles form sedimentary rocks under the water

5. (a) What would cause the surface shape of the land to change very quickly?
    (b) Describe one or two examples where the landscape changed suddenly. If possible, include the date and place of each one. Use your school library to help you.
6. Under what circumstances would rivers alter the landscape the most? Explain your answer.
7. Look carefully at the photographs in Figures 2.12-2.20.
    (a) Which of these types of scenery have you observed?
    (b) Where did you see each type?
    (c) Explain the cause(s) of each type of scenery you have observed.

Making and Reading Maps

Photographs are one way we can use to discover or record what the land is like. Another way is to use maps. Maps are also often used to describe routes from one place to another.

People have been making or reading maps for thousands of years. They may be as simple as a pattern scratched in loose sand with a stick, or as complicated as those used today by engineers.

SCALE

What a mapmaker has to do is to put the important information about a large piece of land onto a small piece of paper. Therefore, the first thing that must be done is to measure the size of the land to be mapped and the size of the paper. Then you can work out how much you must "shrink" the land to make it fit the paper.

"Shrinking" is better explained as the **scale** of the map. The scale of a map is the length used on the map to represent a certain distance on the ground. The scale of a map can be expressed in three ways. One method is called the **line scale.** A line is drawn on the map. It is then divided into units such as kilometres, as shown below.

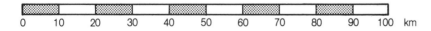

```
0    10   20   30   40   50   60   70   80   90  100  km
```

8. Using the line scale above and a ruler, measure and record the distances in kilometres from (a) A to B (b) B to C (c) A to C and (d) the total distance of ABCD. NOTE: Measure from the dot beside the letter.

C ●

A ●

B ●   D ●

Another frequently used method is the **written scale,** shown like this: 1 cm represents 1 km.

---

9. Using a scale of 1 cm represents 1 km and your ruler, find the distances in kilometres from the following sketch: (a) E to F (b) G to H (c) E to H and (d) EGF.

G ●   F ●

E ●

                                                    H ●

The third method, called a **ratio scale,** is usually shown in a form like this: 1:50 000. This means that 1 cm on the map represents 50 000 cm on the ground. Of the three scales described here, the ratio is used least often.

---

10. Use the given scale of 1:50 000 and your ruler to discover the distance in kilometres of (a) J to K (b) L to M and (c) N to K. NOTE: 100 000 cm = 1 km.

                                                    M ●
        K ●

    J ●

                                                        N ●

            L ●

11. Which of the three scales do you find the easiest to use? Why?
12. *A Map of Your Desk*
    This is a fairly simple map, but you can draw it the way that all maps are drawn. You will need a pencil, a ruler, and a blank sheet of paper.
    (a) Remove all objects from your work desk or table.
    (b) Measure the length and width of your blank sheet of paper. You must somehow "shrink" your desk to fit it onto the sheet of paper. For example, if your desk is 60 cm x 48 cm and your paper is 25 cm x 20 cm, to fit your desk on the paper you should shrink it to 1/3 of its size. You can then have 1 cm on the paper represent 3 cm on the desk. You should then draw your desk 60/3 cm long by 48/3 cm wide on the paper. This means that on the paper your desk becomes 20 cm x 16 cm. Draw your desk top on your paper.

40

(c) There are three ways in which you could have expressed this scale. Using the desk example, the three ways would be 1:3 (a ratio scale); 1 cm represents 3 cm (a written scale); and

0 3 6 9 cm

a line scale. Write your scale, using the three methods, somewhere at the bottom of your map. NOTE: Remember to use *your* figures.

(d) Put three objects on the table. By making careful measurements, transfer them onto your map in their exact positions and reduced in size by the correct amount.

(e) Give a proper title to your map, label all its parts, and put a north arrow on it.

The larger you make a map of a specific area, the more detail you can put on it. Such maps are called **large scale maps.** They only show a small area, but include a large amount of detail about that area. A good example of a large scale map is a map of a town or city, showing all the roads, and sometimes some of the buildings.

Most atlases contain maps that have a small scale. Sometimes even very large cities may be shown by just a dot. Only the smallest amount of detail of an area can be given on a **small scale map.** Each small scale map, however, covers a large area.

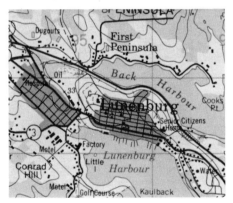

(a) Large scale 1:50 000   (b) Small scale 1:3 150 000

Figure 2-22 A large and a small scale map of part of Nova Scotia

13. Using the two maps in Figure 2.22, do the following:
    (a) Find Lunenburg on each map.
    (b) Which map would you use to find your way around the town of Lunenburg?
    (c) Which map would you use to travel to Halifax?
    (d) Which map would be more useful for planning a hike?
    (e) List the advantages and disadvantages of large and small scale maps.

DIRECTION

A key function of maps is to help people determine a direction for travel. It is often necessary to **orient,** or lay out, the map correctly. The north direction on the map should correspond to true north in the area where the map is being used.

To get your map the right way round, here is what you must do:
- Put your compass on a flat surface.
- The red or black end of the compass needle will point to the north.
- The top of your map is usually north. (On some maps there is a north arrow.)
- Turn your map so that it is lined up as shown in the margin.

The main points of the compass are shown here. There are many others in between also. NOTE: One way to remember direction is with the word "WE"; West is on the left and East is on the right.

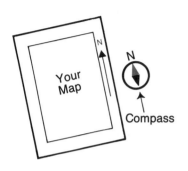

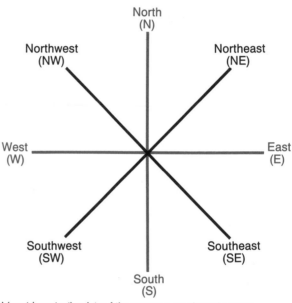

Figure 2-23 The eight major points of the compass

NOTE: The cardinal (most important) points of the compass are shown in colour.

TOPOGRAPHIC MAPS

Topographic maps are fairly detailed maps that include information about the shape of the land. These maps also give information about roads, railways, buildings, vegetation, and many other features. They are a very important "tool" for geography students. Once you learn how to read these maps, you will be able to recognize the shape of the land in any area for which you have a topographic map.

Look at the photograph and map in Figures 2.24 and 2.25 on page 43. The photograph shows how the land appears from an airplane. The lake, trees, fields, and roads are all labelled for you.

Each of the four features you have described from the photograph is shown by a different colour on the map.

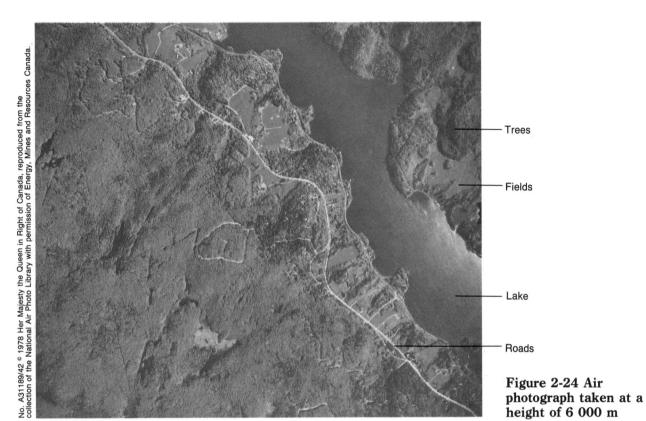

— Trees

— Fields

— Lake

— Roads

Figure 2-24 Air photograph taken at a height of 6 000 m

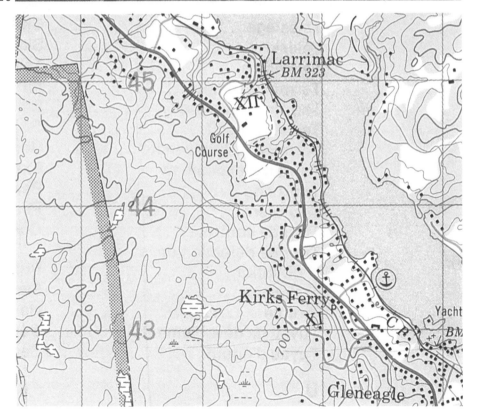

Figure 2-25 Map made from the photograph above

43

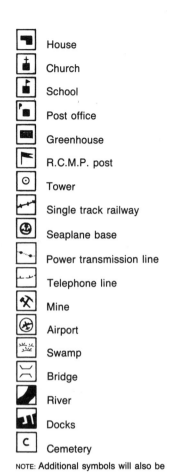

House

Church

School

Post office

Greenhouse

R.C.M.P. post

Tower

Single track railway

Seaplane base

Power transmission line

Telephone line

Mine

Airport

Swamp

Bridge

River

Docks

Cemetery

NOTE: Additional symbols will also be shown on a map legend.

Figure 2-26 Selected map symbols

14. From the photograph, describe the appearance of the following:
    (a) the lake (b) the trees (c) the fields (d) the road
15. Topographic maps are usually shown in colour. Obtain one from your teacher and explain what the following colours represent:
    • blue • green • white • red • pink • black • orange

The other way of showing features on a map is with **symbols.** A symbol is a shape or pattern that represents an object. Symbols and their meanings are shown in a **key** or **legend** on the map. Here are some examples of symbols used in maps.

16. In your notebook, draw each of the symbols that you can see on the map in Figure 2.25, on page 43. Beside each one, write down what that symbol means.
17. Look at the photograph and the map on page 43.
    (a) In what ways do you consider the photograph more useful than the map?
    (b) In what ways do you consider the map more useful than the photograph?
    (c) What job would require the extensive use of air photographs and maps? Explain in what ways they would be used.
    (d) In what specific ways have people benefited as a result of the development of air photographs and topographic maps?

Something else that you have probably noticed on the map are the straight lines. They are shown in blue on a coloured map. These lines are used for pinpointing places on a map. Because this system was worked out by the army, it is called the **military grid.** They worked out this method because they needed an accurate way of describing a location.

Every line on the map has been given a three-figure number. The numbers for pinpointing A, B, and C are given in Figure 2.27. They are called **military grid co-ordinates.**

18. What are the military grid co-ordinates for D and E?
    NOTE: Remember to use the numbers along the top or bottom before using the ones on the side.
19. Why is it especially important for the army to be able to describe locations accurately?

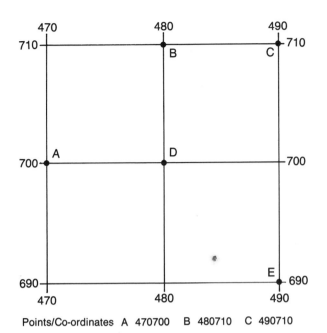

Points/Co-ordinates A 470700 B 480710 C 490710

Figure 2-27 Military grid

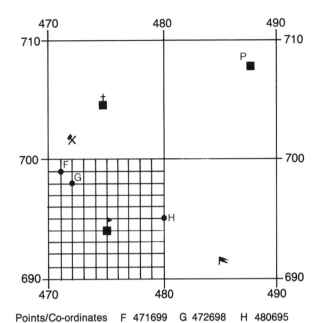

Points/Co-ordinates F 471699 G 472698 H 480695

Figure 2-28 Military grid with subdivisions

Imagine that each of these squares is divided into 100 small squares. In Figure 2.28, one square has been divided up into 100 small squares to show you how it works. These small squares do not appear on maps.

20. **What are the co-ordinates for the church and the school? (See Figure 2.26, page 44, for the symbols.)**
21. **What symbols are found at these co-ordinates?**
    (a) 485692 (b) 488708 (c) 472702
    Explain what each symbol means.

Contour Lines

By looking at the photograph (Figure 2.24) on page 43, you can tell that the land is somewhat uneven. On topographic maps, the height of the land is shown by brown contour lines.

The basis for most modern maps is air photographs. **Cartographers** take these photographs and other information and use them to construct maps and contour lines. Each **contour line** joins points that have the same elevation (height) above sea level.

You will notice in the map that contour lines have been drawn for every 20 m change in elevation. This is called the **contour interval** of the map. Contour intervals vary from one map to another, but are usually the same all over one map.

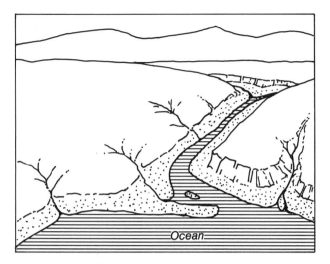

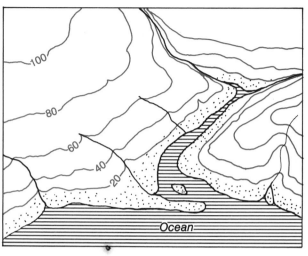

Figure 2-29 Hills and valleys viewed from one side

Figure 2-30 A map of the same area using contour lines to show the shape of the land

22. What is the contour interval of B in Figure 2.31?
23. The model of the land (Figure 2.29) and the map (Figure 2.30) should help you to understand contour lines better.
    (a) The long arm of land going out into the ocean looks very flat. Are there any contour lines on it?
    (b) The hills on the left and the cliffs on the right of the model are very steep. Describe the arrangement and spacing of the contour lines in these places.
24. Finish the following statements:
    (a) When contour lines on a map are close together...
    (b) When contour lines on a map are far apart...
25. Notice where the streams have carved small valleys in the sides of the hills. Find these small streams marked on the map. Describe the pattern made by contour lines in a valley.
26. Imagine that you have been hired by a mysterious multimillionaire who has just bought a whole island off the coast of British Columbia. Your job is to design a paradise for him to live in, on this island. You can have as much money as you need to remodel the island. You must include a large house and a landing strip for the planes that will be coming. Draw the plan for the island on a blank sheet of paper. The scale for your map should be 1 cm represents 0.5 km. These are the details of the island itself, and of your map.

    • The island measures 8 km by 7 km.
    • The highest point is 255 m.
    • There are some beaches and plenty of forests.
    • The contour interval of your map should be 25 m.
    • Include as many map symbols as you can.
    • Include a north arrow, a map title, a legend, and a scale.

    NOTE: Your contour lines should not cross or touch each other.

Contours are useful for discovering the shape of the land and also for drawing profiles of the land, such as those of Canada that you saw earlier.

Six small maps are shown below. Each one represents landscape typical of one of the six regions described earlier, in Figures 1.1 and 1.2 on pages 2 and 3.

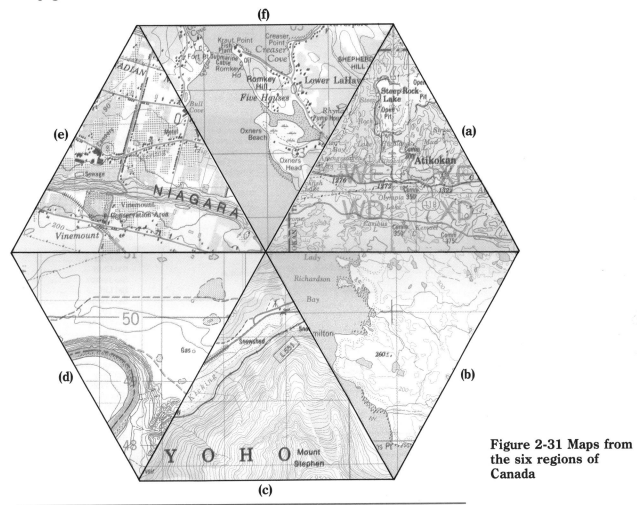

Figure 2-31 Maps from the six regions of Canada

27. (a) Write down the letters given to the two areas that are the flattest.
    (b) Write down the letters given to the two areas which contain the steepest slopes.
    (c) Write a general description of the land and vegetation of the two remaining regions.
    (d) Choose two of the maps and explain how people could make use of the land.
    (e) Imagine that you had a choice of living in any of the six areas shown in the map.
        (i) In which area would you choose to live, and why?
        (ii) In which area would you least like to live, and why?

28. **Construct a table in your notebook like the one below. Fill in the details in the blank boxes. The first line is done for you.**

REGION	LETTER	TWO REASONS
Western Mountains	C	steep, ice in high parts
Interior Plains		
Canadian Shield		
Great Lakes- St. Lawrence Lowlands	**SAMPLE**	**ONLY**
Appalachian Mountains		
Arctic		

A profile may be drawn from any map that has contour lines. To construct a profile of Herman Island, follow these instructions.

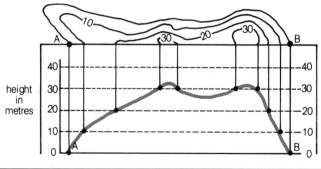

Figure 2-32 Herman Island off the east coast of Nova Scotia

NOTE: Heights are in metres.

(a) The line AB has been drawn across the island. This line will be used for the profile.

(b) Slide a sheet of lined paper along Herman Island until the top of this sheet meets the line AB.

(c) Draw a box on your lined sheet of paper. Label each horizontal (left-to-right) line with one of the numbers of the contour lines of the island.

(d) You *must* use your ruler here. Draw a line carefully straight down from each contour line to the correct line on the graph you drew. Only draw a line from where a contour line meets the top of your lined sheet. For each line, put a dot at the end of your lined sheet of paper.

(e) Join all of the dots with a curved line.

(f) Label the left side of your cross section "Height" and give your diagram a title. Mark A and B on your diagram.

(g) Your profile is finished! That is how Herman Island would look if you cut it in half from A to B.

Figure 2-33 Profile of Herman Island

Climate

"Everybody complains about the weather, but nobody does anything about it." *Mark Twain*

Weather is important to all of us, but for certain people weather conditions are crucial to their lives.

---

29. (a) In your notebook, construct a table with three columns and the following headings:

OCCUPATION	THE IMPORTANCE OF WEATHER TO THE JOB	SPECIAL MEASURES TAKEN TO OVERCOME THESE PROBLEMS
	SAMPLE ONLY	

(b) In the "occupations" column, list the following:

Airline pilot   Construction worker   A person who
Farmer   Telephone line repairer   delivers home-
Firefighter   Supermarket cashier   heating oil
Student   Bus driver

List one other occupation of your choice.

(c) Fill in the remainder of the table. (Assume that each of the people listed above works near to your home.)

Weather results from constant changes that take place in the atmosphere. It includes the day-to-day changes in atmospheric conditions. The **atmosphere** is the layer of air that surrounds the earth.

Climate is the average of weather conditions for a certain place taken over many years. The climate map (Figure 2.36) summarizes the great variety of climates we experience in Canada. Only a few other countries in the world have such great differences in climate.

Figure 2-34 Weather in our atmosphere
"Weather" includes temperature, precipitation, wind, and clouds. It takes place mainly in the lower layers of the atmosphere.

30. Look carefully at the climate map (Figure 2.36) on page 51. Choose two climates that you consider to be the most different from each other. Explain what these differences are.
31. (a) Find Singapore on a map of the world.
    (b) Using a climate map, describe the rainfall and temperature conditions experienced in Singapore.
    (c) Describe how the people's lives in Singapore would vary from yours because of the differences in climate.
    (d) From the point of view of climate alone, would you prefer to live in Singapore or where you live now? Explain your answer.
32. Using the climate map on page 51, do the following:
    (a) Rank in order, the names of the cities and towns marked on the climate map. Put the one with the lowest winter temperature first. Beside each place, write the name of the province or territory in which it is located, and its winter temperature.
    (b) Using a few sentences, describe the variation of temperatures in Canada in the winter.
    (c) In a short paragraph, describe the variation of temperatures in Canada in the summer.

Precipitation, which includes rain and snow, varies throughout the country as much as temperature. Figure 2.35 shows the distribution of annual (yearly) precipitation, averaged out over 35 years.

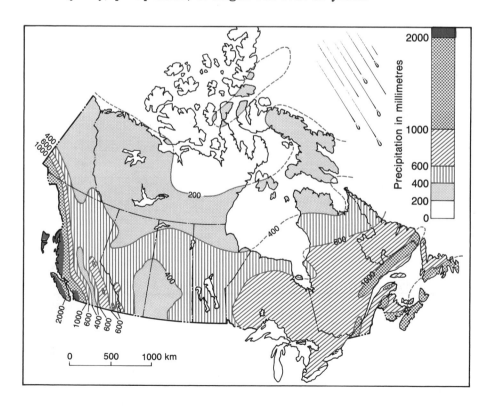

Figure 2-35 Average annual precipitation

50

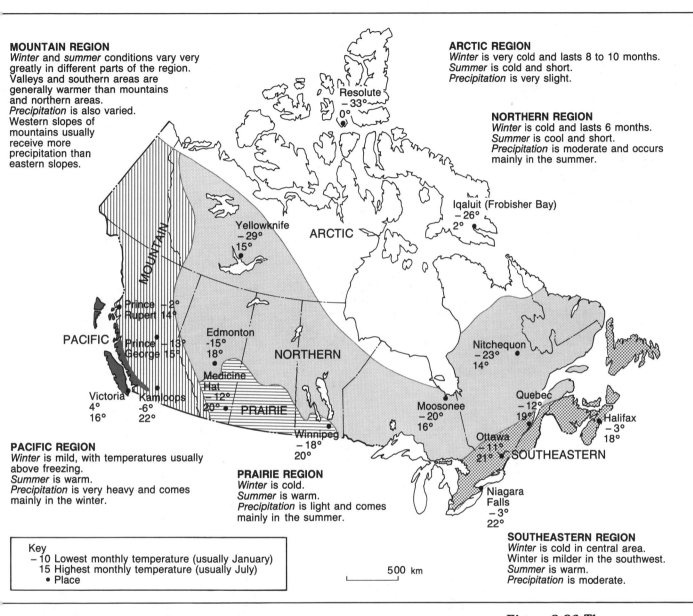

MOUNTAIN REGION
Winter and *summer* conditions vary very greatly in different parts of the region. Valleys and southern areas are generally warmer than mountains and northern areas. *Precipitation* is also varied. Western slopes of mountains usually receive more precipitation than eastern slopes.

ARCTIC REGION
Winter is very cold and lasts 8 to 10 months. *Summer* is cold and short. *Precipitation* is very slight.

NORTHERN REGION
Winter is cold and lasts 6 months. *Summer* is cool and short. *Precipitation* is moderate and occurs mainly in the summer.

Resolute
−33°
0°

Yellowknife
−29°
15°

ARCTIC

Iqaluit (Frobisher Bay)
−26°
2°

MOUNTAIN

Prince −2°
Rupert 14°

PACIFIC

Prince −13°
George 15°

Edmonton
-15°
18°

NORTHERN

Nitchequon
−23°
14°

Medicine
Hat
−12°
20°

PRAIRIE

Victoria   Kamloops
4°   −6°
16°   22°

Winnipeg
−18°
20°

Moosonee
−20°
16°

Quebec
−12°
19°

Halifax
−3°
18°

Ottawa
−11°
21° SOUTHEASTERN

Niagara
Falls
−3°
22°

PACIFIC REGION
Winter is mild, with temperatures usually above freezing. *Summer* is warm. *Precipitation* is very heavy and comes mainly in the winter.

PRAIRIE REGION
Winter is cold. *Summer* is warm. *Precipitation* is light and comes mainly in the summer.

SOUTHEASTERN REGION
Winter is cold in central area. Winter is milder in the southwest. *Summer* is warm. *Precipitation* is moderate.

Key
−10 Lowest monthly temperature (usually January)
15 Highest monthly temperature (usually July)
• Place

500 km

Figure 2-36 The climatic regions of Canada

33. Study the map of precipitation. Copy the following paragraph into your notebook. Each asterisk (*) represents a missing word. Fill in the blanks in your notebook.

The wettest area of Canada extends in a narrow strip along the * coast. The other wet areas are along the * coast and in some isolated areas of **. The part of Canada that receives the least precipitation is the central *. Dry areas are found in much of the * mainland, the * parts of Alberta and Saskatchewan, and some parts of ** . The remaining areas of Canada have a moderate precipitation.

USING CLIMATIC DATA

Every day, in over two thousand places throughout Canada, people take measurements of the weather.

Inside the screen shown in Figure 2.37, special thermometers record temperatures. Other recording instruments may be placed inside the screen to measure other factors, such as humidity.

The weather balloon shown in Figure 2.38 is filled with helium. It carries instruments that send messages about temperature and humidity. It is also used to measure wind speed and wind direction at different heights.

Weather measurements are sent to nine Regional Centres across Canada. Information is automatically received by a computer and is used to make forecasts.

Figure 2-37 Observations of temperature are made in a white Stevenson screen *This Stevenson screen contains thermometers to record maximum, minimum, and present temperatures. It also has wet and dry bulb thermometers which are used to determine relative humidity.*

 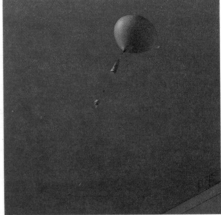

Figure 2-38 Preparing a weather balloon to take high-level measurements *The balloon filled with helium is released and carries a radiosonde aloft. The radiosonde transmits information about temperature, humidity, and pressure as it rises. The balloon's course is tracked automatically so that high level winds can be plotted.*

Figure 2-39 (a) The computer which receives and stores weather information

Figure 2-39 (b) The forecaster can call up information on the display terminal at each desk

The information is coded by the observer in a special sequence of letters and numbers. Each has a particular position and meaning.

Initially, the information is fed into the computer and, using information from all over North America, the Pacific and Atlantic Oceans, the meteorologists draw weather maps. Meteorologists are then able to make forecasts based on these maps.

```
TOL SA 2050 W1 X 1/8L-F 069/34/34/0506/971/R07VR10-/ 71200
FDY SA 2054 W2 X 1/4F 064/40/40/0000/970/ 614
DAY SP 2126 M11 OVC 21/2F 1806/969
CVG SA 2050 M13 OVC 7 061/57/54/1810/971/ 717 15//
CMH SA 2050 M10 OVC 4F 076/51/47/1807/973/ 712 16//
MFD RS 2050 M4 OVC 21/2F 073/49/46/1811/973/ 72100 16//
ZZV SA 2054 E7 OVC 5F 085/49/48/1910/977/ 610
CAK SP 2139 M5 OVC 4F 1809/975
CLE SP 2137 M9 OVC 11/2RW-F 1910/973
YNG SP 2139 M7 OVC 5F 1707/975
PIT SA 2050 M4 BKN 10 OVC 11/2F 095/46/43/1904/979/SFC VSBY 3
    LB18E37 CIG RGD/ 70700 16//
AGC SP 2115 W0 X 1/8L-F 2009/979/R28VR06-
DUJ SA 2050 W2X 1/8F 106/36/35/1405/980/LE11
PSB RS 2055 W3 X1F 128/33/31/1506/985/ 607
ERI SP 2130 M13 OVC 21/2L-F 3405/975
JHW SA 2045 W0 X 0L-F 40/40/1805/973
IAG SA 2055 W1 X 1/4L-F 092/34/32/0000/978/RVRNO/ 302 FSR12F WET=
BUF SA 2050 W1X 1/8L-F 095/36/32/0203/978/R23VR06V08/ 30300
ROC SA 2053 -X M12 OVC 4F 102/38/35/1403/981/F1/ 61201 16//
ELM SA 2055 M18 OVC 7 119/36/32/0904/985/ 607
BGM SA 2050 M10 OVC 3F 121/34/32/1707/983/ 607 16//
SYR SA 2050 M20 OVC 4F 115/32/31/0810/985/ 60801 15// 90408
UCA RS 2100 -X 3 SCT M11 BKN 15 OVC 2ZL-F 131/28/28/1015G20/988/
    630
RME SA 2056 -X 5 SCT M11V OVC 2ZL-F 130/28/28/1108/989/LF2 CIG
    9V12 WND 08V14 CIG RGD/ 608 15// IR13=
ART SA 2055 E18 OVC 3ZL-F 115/28/26/0108/986 610
MSS SA 2050 -X E25 OVC 21/2IP-F 154/19/17/0914/997/F4/SEIPB44/ 717
PBG SP 2112 W10 X 2IP-S-F 0000/000=
```

Figure 2-39 (c) The encoded information in printed form *The first line of the code means the following: Toledo, a regular report, 2050 hours GMT, indefinite ceiling, 100 feet high, obscured, visible horizon 1/8 mile, drizzle and fog, air pressure 1006.9 mb, temperature 34°F, dewpoint 34°F, wind from 050° at 6 knots, altimeter setting 29.1'', runway visibility 0710', pressure has fallen 1.2 mb in the last 3 hours. Note: Information from the U.S.A. comes in non-metric form. In Canada information is in metric units.*

Figure 2-40 Weather maps being plotted *Using maps which show information about different layers of the atmosphere, satellite photographs, and radar information, the meteorologist is able to compile maps which help in the creation of weather forecasts.*

Figures recording temperature and precipitation for any one place are often summarized in the following way.

MONTHS	J	F	M	A	M	J	J	A	S	O	N	D	YEAR
Monthly Mean Temp (°C)	-6.7	-7.2	-3.2	2.3	8.6	14.1	18.4	17.9	13.9	8.6	3.3	-3.6	5.6
Monthly Mean Total (mm)	98	82	76	75	80	79	74	90	92	99	115	100	1060

NOTE: Temperature information is given on the top line of figures, with precipitation values for each month below it.

Figure 2-41 Climatic data for Charlottetown, P.E.I.

These figures can be plotted as a climograph. A climograph uses a line graph for temperature and a bar graph for precipitation. In Charlottetown, the maximum (highest) average monthly temperature of 18.4°C was recorded in July. The minimum (lowest) average monthly temperature of −7.2°C was recorded in February. The annual range of temperature in Charlottetown is 18.4 − (−7.2)°C which is 25.6°C. Thus "range" is the difference between the highest and lowest temperature at a particular place.

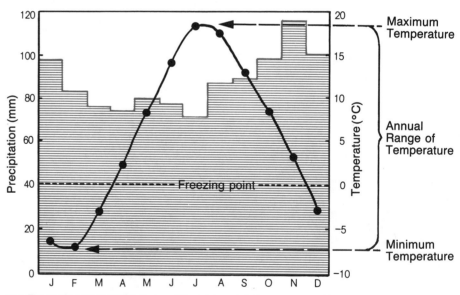

Figure 2-42 Climograph for Charlottetown, P.E.I.

NOTE: The yearly average temperature and the total precipitation are not graphed.

You can observe that precipitation in Charlottetown is distributed fairly evenly throughout the year but is highest during the winter months. By looking at the temperatures between the months of December and March, you will be able to see that much of this precipitation will come in the form of snow.

Figure 2.43 shows climatic data for several places in Canada.

LOCATION Altitude, Latitude & Longitude	J	F	M	A	M	J	J	A	S	O	N	D	Year
Resolute, N.W.T. 64m 75°N 95°W	-32.6	-33.5	-31.3	-23.1	-10.3	-0.3	4.3	2.7	-4.9	-14.7	-24.1	-28.2	-16.4
	3	3	3	6	9	12	26	30	18	15	6	5	136
Whitehorse, Y.T. 698m 61°N 135°W	-18.9	-13.2	-7.7	-0.1	7.1	12.4	14.1	12.3	7.8	0.7	-9.0	-15.8	-0.8
	18	14	15	11	14	29	33	36	29	20	22	20	261
Vancouver, B.C. 5m 49°N 123°W	2.4	4.4	5.8	8.9	12.4	15.3	17.4	17.1	14.2	10.1	6.1	3.8	9.8
	147	147	117	61	48	45	30	37	61	122	141	165	1121
Calgary, Alta. 1079m 51°N 114°W	-10.9	-7.4	-4.3	3.3	9.3	13.2	16.5	15.2	10.7	5.7	-2.6	-7.6	3.4
	17	20	20	30	50	92	68	56	35	19	16	15	438
Churchill, Man. 35m 69°N 94°W	-27.6	-26.7	-20.3	-11.0	-2.3	6.1	12.0	11.4	5.7	-1.0	-11.9	-21.8	-7.2
	14	13	18	24	28	40	49	58	52	40	40	20	396
Toronto, Ont. 116m 43°N 79°W	-4.4	-3.8	0.6	7.6	13.2	19.2	21.7	21.1	17.0	11.2	4.8	-1.8	8.9
	62	57	66	67	73	63	81	67	61	62	67	64	790
Goose Bay, Nfld. 44m 53°N 60°W	-16.3	-14.4	-8.4	-1.8	4.9	11.1	15.8	14.5	9.8	3.2	-3.6	-12.3	0.2
	69	60	69	54	62	82	102	93	76	72	70	68	877
Halifax, N.S. 41m 45°N 63°W	-3.8	-4.2	-0.7	4.0	9.0	13.7	17.6	17.8	14.8	9.9	4.9	-1.1	6.8
	147	128	112	105	110	85	92	94	94	113	152	148	1380

Figure 2-43 Climatic data for selected locations

34. (a) During which month do most locations record their maximum temperatures?
    (b) During which month do most locations record their minimum temperatures?
35. (a) Plot climographs from two places in Canada that are far apart. Put titles and *all* necessary information on your graphs.
    (b) Compare and contrast (describe the differences and similarities) the temperature and precipitation of the two places you have chosen to plot in 35 (a). Include in your account the maximum, minimum, and range of the stations, and the precipitation totals.

THE FACTORS THAT DETERMINE CLIMATE

The following are four main factors that determine climate:
(1) latitude
(2) wind direction and the position of water bodies
(3) mountain ranges and altitude
(4) ocean currents

Latitude

The land area of Canada extends from 42°N to 83°N. This affects temperature for two specific reasons. See Figure 2.44.

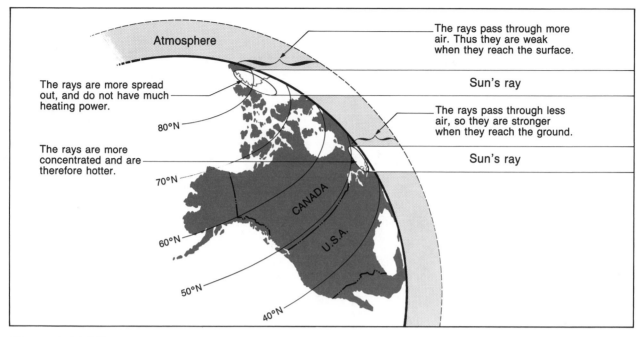

Atmosphere

The rays pass through more air. Thus they are weak when they reach the surface.

Sun's ray

The rays are more spread out, and do not have much heating power.

80°N

The rays pass through less air, so they are stronger when they reach the ground.

Sun's ray

The rays are more concentrated and are therefore hotter.

70°N

CANADA

U.S.A.

60°N

50°N

40°N

Figure 2-44 Why latitude affects climate

36. (a) Conduct the following experiment. Write down your observations.
   Purpose: To demonstrate the spread of the sun's rays on different parts of the earth's surface.
   Equipment: Flashlight with a strong beam, globe, chalk, and a darkened room.
   Method: You will need three people in each group. One person holds the globe upright. The second person holds the flashlight about three metres from the globe. The third person uses the chalk to mark the rings of light onto the globe.
   Darken the room and, making sure to hold the flashlight at the same height, shine the light on the following:
   (i) the Equator   (ii) southern Canada   (iii) the Arctic
   (b) Explain the significance of your results to temperatures on the earth's surface.
37. (a) For this exercise, use information that you can get from the map of climatic regions (Figure 2.36 on page 51). Choose four places whose differences support the statement that latitude is important in controlling temperature. For each place, list the name, the latitude, and the January and July temperatures.
   (b) Within Canada, what influence does latitude have on the following:
   (i) the ease of transporting goods
   (ii) the choice of an area in which to live
   (iii) the production of food
   (iv) the types of houses that are built

Wind Direction and the Presence of Water Bodies

Bodies of water, especially large oceans, provide most of the moisture needed to create precipitation. The winds carry this moisture onto the land, where it may fall as rain or snow.

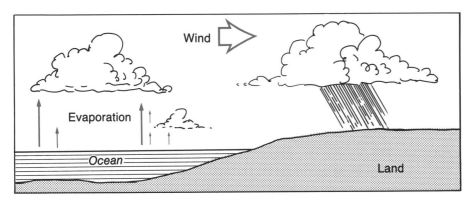

Figure 2-45 Most of our moisture originates from large oceans

Water is also able to store a great quantity of heat. The water warms up slowly in the early summer, and cools the land close to it.

Water stores heat during the summer and gives it up slowly during winter. This keeps the nearby land warmer than inland locations.

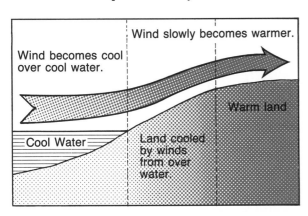

Figure 2-46 The cooling effect of the oceans in early summer

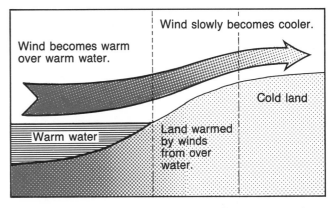

Figure 2-47 The warming effect of the oceans in early winter

38. Temperatures recorded in Victoria, B.C., Medicine Hat, Alta., Winnipeg, Man., and Halifax, N.S. are shown below.

LOCATION	MINIMUM TEMPERATURE °C	MAXIMUM TEMPERATURE °C
Victoria, B.C.	4	16
Medicine Hat, Alta	− 12	20
Winnipeg, Man.	—18	20
Halifax, N.S.	− 3	18

Group the cities into pairs based on similarities of their summer and winter temperatures.

39. Using the figures from the table, answer these questions.
    (a) What effect does the proximity (or closeness) of the ocean have on (i) winter temperatures and (ii) summer temperatures?
    (b) The prevailing (most frequent) wind in southern Canada is the westerly wind, since it blows from the west. Use this fact to explain why Victoria is warmer in the winter than Halifax.

Mountain Ranges and Altitude

As all mountain climbers know, temperatures decrease as they approach the summit (top). On average, temperatures drop by 0.6°C for every 100 m increase in altitude (height above sea level).

A line of mountains, and even an individual mountain, has an effect on precipitation. More rain or snow falls on the windward side than on the leeward side. The drier leeward side is also called the "rainshadow." The reasons for this will be explained in the section on orographic precipitation on page 60.

40. In which part of Canada would you expect to find altitude playing an important part in determining temperatures and precipitation patterns?
41. (a) Mount Logan, in the western Yukon, is the highest mountain in Canada with its summit at 6 050 m.
       (i) If temperatures decrease at a rate of 0.6°C/100 m, work out how much cooler it is at the summit than at sea level.

Figure 2-48 The effect of altitude on temperature

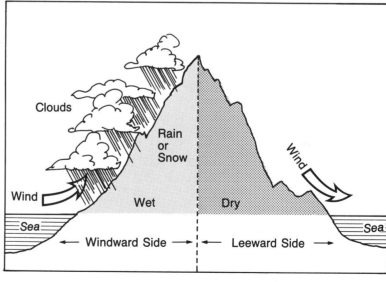

Figure 2-49 The effect of a mountain range on precipitation

(ii) When the temperature at sea level is 20°C, what is the temperature at the summit?

(iii) If the temperature at sea level is 24°C, at what height would the precipitation be in the form of snow?

(b) Find Mount Logan on an atlas map. What effect has altitude had on the scenery of this area?

Ocean Currents

The movement of water in the oceans may bring warm water to cooler areas or cool water to warm areas.

Warm water gives off more moisture than cool water. Therefore ocean currents may also affect the amount of precipitation received by the nearby land.

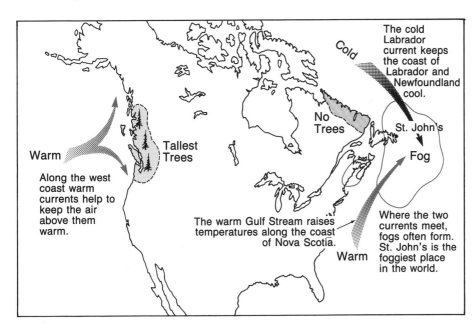

Figure 2-50 Ocean currents around Canada's coastlines

42. All of the four climatic factors we have covered have some influence on Canada's climate.

(a) Which two factors do you think have the greatest effect on the area where you live? Give the reasons for your answer.

(b) Which factor do you think would have the least effect in your area? Explain why you chose this factor.

PRECIPITATION

Water exists in three states—solid, liquid, and gas. The changes from one state to another are illustrated in Figure 2.51 on page 60.

Figure 2-51 The states of water in the atmosphere

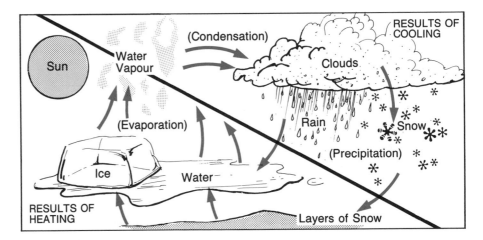

Heat energy is the key factor in allowing water to change from one state to another. For example, **water vapour,** an invisible gas in the air, is formed when heat is applied to ice, snow, or liquid water. This effect is seen when the sun shines and water in rain puddles **evaporates** to become water vapour. When water vapour cools, it can **condense** to become liquid water once again. In some cases, when water vapour condenses it forms tiny water droplets in the air that may mass together to form clouds.

As you can see from Figure 2.52, air rises and cools, and the water vapour in the air condenses to form clouds. Under suitable conditions, precipitation will fall from the clouds.

There are three main ways in which the air is forced to rise. The types of precipitation that result are called (1) orographic (or relief) precipitation, (2) convectional precipitation, and (3) cyclonic (or frontal) precipitation.

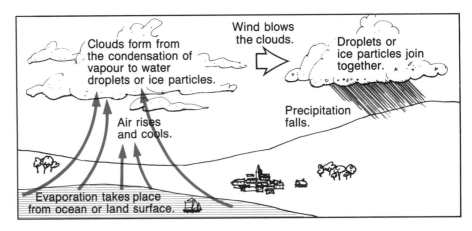

Figure 2-52 The stages that lead to clouds and precipitation

Orographic Precipitation

In all parts of Canada a rise of land may cause clouds to form and **orographic precipitation** to occur. However, this is most marked along the Pacific coast and Western Mountains.

As you can see from Figure 2.54 and from the precipitation map (Figure 2.35, page 50), precipitation occurs on the windward side of each mountain range, and is heaviest along the coast. On the leeward side of the mountains, especially in western Alberta, the Chinook winds occur fairly frequently. They may be beneficial in early spring when they help to clear the snow from the land so that the cattle can graze. Later in the year, however, they may dry the soil, causing grass and crops to wither.

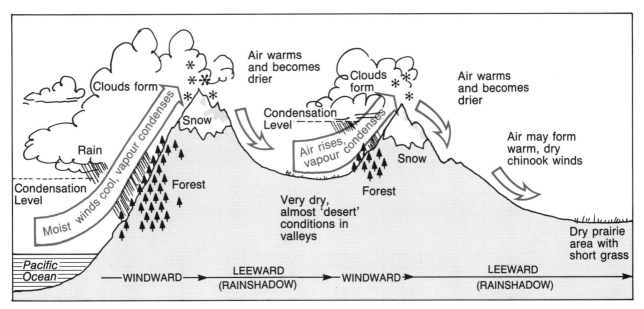

Figure 2-54 Orographic precipitation and its effects in western Canada

Convectional Precipitation

In the summer months many hailstorms and thunderstorms occur over hot land surfaces. These may be very violent and cause much damage to ripening wheat or other crops. The air above a heated surface becomes very warm. Because warm air is fairly light, it rises. This rising leads to the creation of clouds from which **convectional precipitation** may occur.

Figure 2-55 Convectional hailstorm in the Prairies

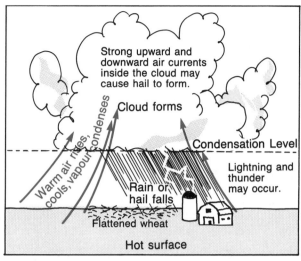

Figure 2-56 Convectional precipitation and its effects in the Prairie Provinces

Cyclonic Precipitation

Canada is located in a part of the world where several different air masses may meet. When warm and cold air meet, the warm air rises above the cold air. Clouds form and **cyclonic precipitation** may follow. Precipitation may be heavy or light, depending on the conditions at that time.

All three causes of precipitation are present to some extent in all regions of Canada, except the Arctic. Certain variations, however, should be considered.

- The Western Mountains are dominated by orographic precipitation.
- The Prairies experience more convectional rainfall than other areas.
- Central and eastern Canada experience much cyclonic rainfall.

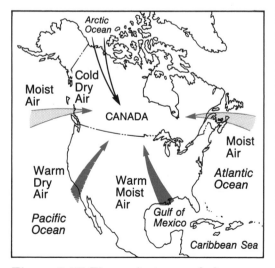

Figure 2-57 The main types of air affecting Canada

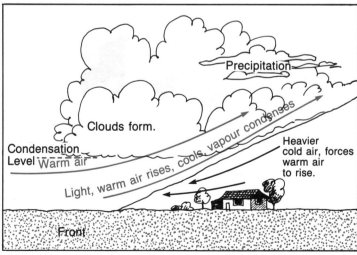

Figure 2-58 Cyclonic (frontal) precipitation

The Canadian Arctic is one of the largest deserts in the world. It is a desert, since it receives less than 250 mm of precipitation per year. This dry nature of the Arctic is due to the absence of those factors that cause precipitation. The absent factors can be summarized in four statements.

- Much of the Arctic is flat land.
- Little warm air reaches this region.
- The land surface does not get very warm.
- The Arctic Ocean is cool in summer and frozen in winter.

43. Using Figure 2.51 on page 60 as a guide, list the stages involved in the transformation of water to snow.

44. (a) Set up the following experiment at school or at home.

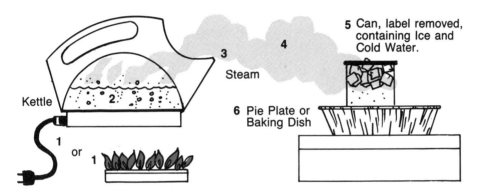

(b) Copy the diagram into your notebook.
(c) Explain what happened at numbers 5 and 6.
(d) Relate each numbered part on the diagram, in order, to one part of Figure 2.51, page 60.

45. (a) Use the precipitation map (Figure 2.35 on page 50) to determine the approximate amount received in your area in an average year. (If you are able to get hold of more accurate information, use it to answer this question.)
(b) Which of the three causes of precipitation do you consider to be most important in your locality? Explain the reasons for your answer.
(c) In what ways has precipitation, or the lack of it, caused problems in your area?
(Flooding, hail damage, traffic disruptions, droughts, etc. are suitable examples. You may be able to think of others.)
In each case explain
(i) what occurred   (iii) what resulted.
(ii) when it happened
(d) Millions of dollars are being spent on experiments to determine how to create rain, prevent hail damage, improve climatic conditions, and divert destructive storms. Do you consider that this money is being used wisely? Explain your answer.

Vegetation, Soils, and Animals

Climate directly and indirectly determines the types of vegetation, soils, and animal life which inhabit an area.

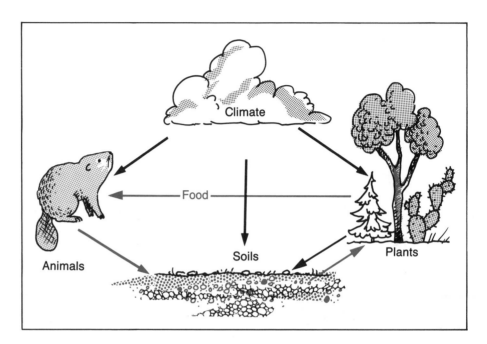

Figure 2-59 The relationships between climate, animals, plants, and soils

---

46. Using the diagram as a guide, give two Canadian examples to show the following:
    (a) Climate directly determines the type of vegetation that grows. Explain the reasons for your choices.
    (b) Climate affects the variety of animals found in an area. Explain your answer.

VEGETATION

Canada's vegetation is a mixture between the **natural vegetation** and the vegetation which has resulted because of people's interference with nature. Natural vegetation is the mixture of plants that grows without human interference. A map showing the major vegetation zones is shown in Figure 2.60.

There are very few parts of southern Canada where the vegetation is natural. This is because people have cleared the land, mainly for use as farms. Even when farms are abandoned, it takes many years for the vegetation to get back to its true natural state.

The most important factor that controls our natural vegetation is the climate. A plant makes its food by using sun, air, and moisture. Different plants have their own special needs and can only grow where these needs are met.

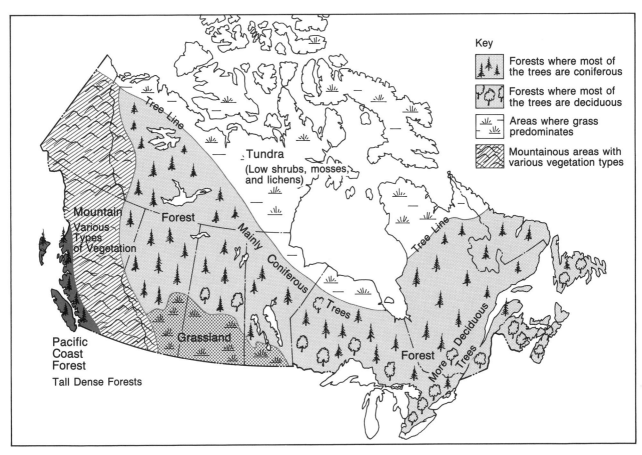

Key

- Forests where most of the trees are coniferous
- Forests where most of the trees are deciduous
- Areas where grass predominates
- Mountainous areas with various vegetation types

Tree Line

Tundra
(Low shrubs, mosses, and lichens)

Mountain
Various
Types
of Vegetation

Forest

Mainly Coniferous Trees

Tree Line

Pacific
Coast
Forest

Tall Dense Forests

Grassland

More Deciduous Trees

Forest

Figure 2-60 Natural vegetation regions of Canada

Tundra

In the Arctic, the winters are long and cold. There are only a few weeks when it is warm. For this reason **tundra vegetation** includes a variety of plants that grow very quickly, flower, and seed during the short summer season. Mosses and lichens are common.

Figure 2-61 Tundra vegetation

Plants that grow in the Arctic are all very short. Plants that are short absorb as much heat as possible from the ground surface. Even though very little rain or snow falls in the Arctic, it is so cool that the ground is quite damp and there is enough moisture for the plants to grow.

Farther south the summers get longer and warmer and the amount of moisture increases. These improved conditions allow more plants to grow. They can also grow much larger.

The Forests

At the southern edge of the tundra vegetation is the **tree line.** South of this line, trees can be found growing over most of the surface. North of this line, there are few trees. The tree line is usually marked on vegetation maps, such as in Figure 2.60 on page 65. The following diagram and photograph give some idea of how the trees in the north become smaller as climate becomes more severe.

At the tree line and for a considerable distance south, most of the trees are evergreens or **conifers,** such as spruce or pine.

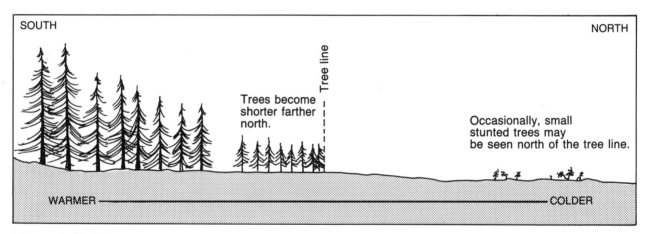

Figure 2-62 The tree line

Figure 2-63 The northern edge of the forest near Moosonee, Ontario

Conifers are cone-bearing trees with thin, needle-shaped leaves. They can stand the cold conditions because of their thick bark, which stops them from losing much moisture. Their needles also contain a material that acts as an anti-freeze during the long cold winters of the north.

Figure 2-64 Coniferous forest

Farther south, the forests gradually change. There are more deciduous trees, such as maple and oak, and fewer coniferous trees. A deciduous tree is one that has bigger, flat leaves, which it loses in the autumn. In the winter it sends its sap down to be stored in its roots. In this way it survives the winter.

Figure 2-65 A pine tree with a close-up of its cones

Figure 2-66 Deciduous trees

Along the west coast of Canada, as you have discovered, the climate is very moist. Huge trees are able to grow. The most famous trees along the west coast are the Douglas firs. These Douglas firs are also the largest trees found in Canada.

Most of the trees in Canada, except the poorer and smaller ones of the far north, provide very valuable timber.

Figure 2-67 Douglas firs on the slopes of Capilano Canyon, British Columbia

Figure 2-68 Grass recovering from a dry spell *A healthy grass plant may have roots that measure up to 500 km in total length.*

The Grasslands

In the southern Prairies, the climate is very dry. There is not enough water for trees to survive without special help. Grass is able to survive because it requires less water than trees, and it has long roots to reach water deep in the soil. Even if the leaves are burned off by the hot sun, the plant will grow again with new fresh blades of grass when it rains.

The Mountains

As you noted earlier, temperatures decrease towards the summit of a mountain, and soils usually get thinner and rockier. As a result, the vegetation is different on various parts of the mountain.

Figure 2-69 (a) Zones of vegetation on a mountain

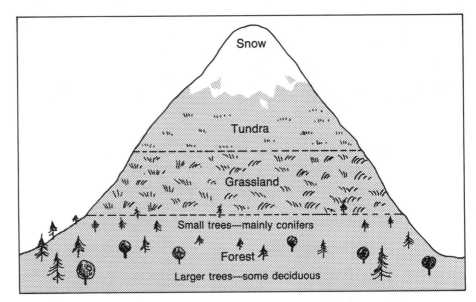

Figure 2-69 (b) A mountain with different kinds of vegetation

68

As you have already learned, many of the mountain valleys are very dry. Unless the valleys are irrigated (watered by people), there may be tough grasses and even some desert-like plants growing there.

Figure 2-70 Vegetation in an irrigated mountain valley

47. Make a collection of ten vegetation samples (such as leaves). Try to find at least one from each of the following groups: coniferous trees, deciduous trees, grass, moss, and lichen.
    In your notebook, make a table with the following headings:

48.

NAME OF VEGE-TATION SAMPLE	WHERE FOUND	THE NATURAL VEGETATION ZONE IN WHICH IT WOULD BE MOST COMMON
		SAMPLE ONLY

49. Using the information about your ten samples, and the vegetation map (Figure 2.60) on page 65, fill in the table.
    (a) Using suitable paper, trace the natural vegetation map from Figure 2.60 on page 65.
    (b) Place your traced map of vegetation over the climate map (Figure 2.36) on page 51.
        (i) Describe in what ways the two maps are similar.
        (ii) Try to explain the reasons for the similarities. Use some examples.

SOIL

If you were to take a handful of soil and look at it with a magnifying glass, you would find that it is made up of five main types of things:
• Grains of sand, silt, or clay, called the **mineral matter.**
• Pieces of decaying plants and dead animals, called **humus.**
• Small living animals and plants.
• Water.
• Air between the particles.

Figure 2-71 How a soil is formed

The rocks are broken into small pieces by weathering and erosion. The pieces either stay on top of the rock or are washed into a valley, where they collect together.

Plants start to grow on the loose rock particles, small ones first. Their roots help to break up the rocks more. Dying roots and leaves collect in and on the young soil.

Small animals, especially earthworms, mix the vegetable matter in with the mineral matter.

The rain washes dissolved materials down through the soil. The sun draws them up towards the surface again. Eventually, layers form in the soil.

It takes hundreds of thousands of years for soil to form. By following the flow chart (Figure 2.71), you will see how soil is created.

If you were to dig down into the soil in most parts of Canada, you would find different layers. These layers make up a **soil profile.** In Canada there are many different kinds of soil. An atlas map will show the distribution of these soil types. Most of Canada has **podzolic soils,** which have a pale layer a few centimetres below the surface. The richest soils in Canada are found in the grasslands of the southern Prairies. These soils are dark, rich, and deep, and the best ones are called **black soils.**

---

50. Find a place where the layers of soil are exposed. You might see this in a road cut, construction site, or river valley.
    (a) Draw a soil profile. For a guide, look at the soil profiles in Figure 2.72. Include a scale to show the depth, 0 cm at the soil surface.

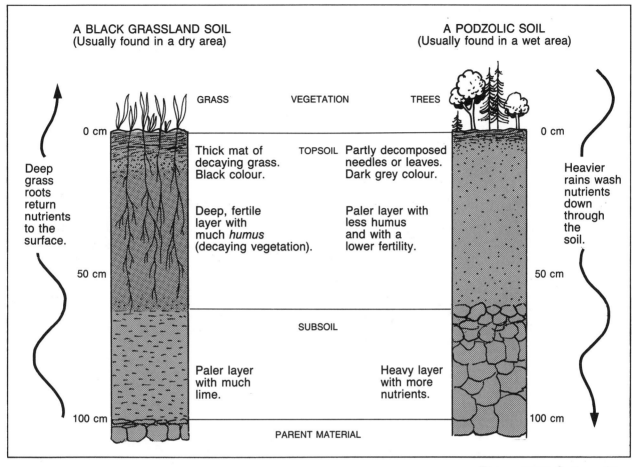

A BLACK GRASSLAND SOIL
(Usually found in a dry area)

A PODZOLIC SOIL
(Usually found in a wet area)

GRASS   VEGETATION   TREES

0 cm

Deep grass roots return nutrients to the surface.

TOPSOIL

Thick mat of decaying grass. Black colour.

Partly decomposed needles or leaves. Dark grey colour.

Deep, fertile layer with much *humus* (decaying vegetation).

Paler layer with less humus and with a lower fertility.

Heavier rains wash nutrients down through the soil.

50 cm

SUBSOIL

Paler layer with much lime.

Heavy layer with more nutrients.

100 cm

PARENT MATERIAL

0 cm

50 cm

100 cm

Figure 2-72 Soil profiles

(b) Neatly colour and describe the different layers in your profile.

(c) Draw and label the type(s) of vegetation that grow(s) on the surface. (This does not have to be to scale.)

51. Copy the following paragraph into your notebook, leaving a gap for a word wherever there is an asterisk (*). Insert one of the following words wherever it is suitable. (One word fits in each gap.)

dark • vary • fertile • Prairies • plants • downward • humus

    Soils in Canada are best in the southern * where it is fairly dry. The dead grass forms * in the soil. This gives it a * colour and helps to keep it *. In most other parts of Canada, soils are poorer because the rain, trickling through the soil, washes nutrients *. In the Arctic areas the soils are thin and stony. They are not very fertile because the * do not supply enough humus. Along the west coast and mountains soils * greatly from place to place.

ANIMALS

Most Canadians live in areas where people's activities have had a great effect on the wildlife. Through our persistence in developing cities and towns, cutting forests, mining, and polluting our air and water, we have probably had some effect on all animal life.

Some animals have managed to survive despite this human interference. Others, however, are in danger of dying out, and several **species** (types) are already **extinct** (no longer in existence). In our cities and towns many smaller wild animals have changed from their more natural ways of life and seem to thrive in the company of people.

Figure 2-73 Urban wildlife

52. For a period of one week, keep a list of the wild animals (including birds, fish, and insects) that you see around your home or school. For each one write down the following information:
    • the type of animal
    • where it lives
    • what it eats
    • what it would eat in a natural environment
53. (a) Do you consider the "wild" urban animals and birds to be an asset (benefit) or a detriment (harm) to your environment? Explain your answer.
    (b) Do your parents and grandparents have the same or different opinions? Explain your answer.

Patterns of Wildlife

The maps (Figure 2.74) attempt to show the distribution of four selected forms of wildlife. Each animal's **habitat** (surroundings it lives in) must provide it with the food supply and climate that it needs to exist.

(a) Moose

(b) Pronghorn Antelope

Figure 2-74 The distribution of four selected forms of wildlife

(c) Polar Bear

(d) Ermine

54. Describe the factors that have determined the distribution of each of the animals included in the maps.

The supply of these necessities may vary from one season to the next, and some animals **migrate** (travel) to areas that will supply their needs.

Migration is a major task. For some animals, their ability to travel great distances fairly easily permits migration over thousands of kilometres. For land animals migration is usually restricted to a few hundred kilometres. Some animals, such as chipmunks, **hibernate** (go to sleep) during the winter months, avoiding the necessity of finding food for this period.

Animals usually follow a regular migration route. For example, birds passing over North America travel along three main "flyway" routes.

As birds follow these major flyway routes, they must stop over to rest and feed in swamps and wilderness areas. Canada's population, however, continues to grow, and mankind's demand for more land increases. As cities expand and more land is used for such activities as farming, mining, and lumbering, there is less land available for our wildlife.

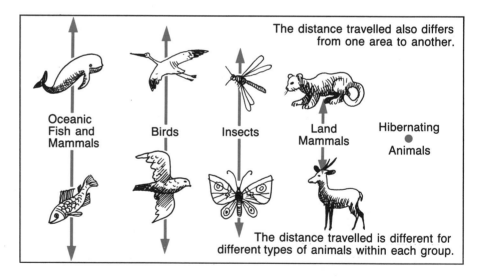

Figure 2-75 The relative distance of migration of various animal groups

55. What physical or climatic factors would influence migratory routes used by
    (a) birds?
    (b) land animals?
56. Using information from your school or local library, study one animal from the list below. Find out where it lives in the summer and in the winter. Mark these places on a map together with the migration route that it follows. Describe its way of life, including its food and breeding habits.

Arctic Tern	North American Monarch Butterfly
Black Poll	Blue Whale
Bobolink	Fin Whale
Golden Plover	Harp Seal
Arctic Fox	Northern Sea Lion
Caribou	Atlantic Salmon
Polar Bear	One kind of Pacific Salmon

57. Should certain portions of land be set aside for migrating birds, even though that land may be needed for other purposes? Explain your answer fully.

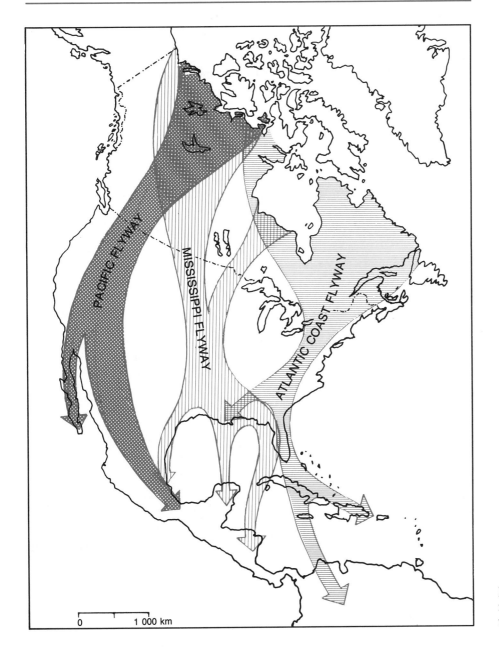

Figure 2-76 The three main overland flyways used by migrating birds

Extinction of Wildlife

Many forms of wildlife that once lived in great numbers in North America no longer exist. Among these is the Passenger Pigeon. During their annual migrations, millions of these birds used to darken the skies for kilometres. People slaughtered them to provide an easy source of food. Other animals that were very nearly eliminated by human "progress" include several varieties of waterfowl, the beaver, the plains bison, and the muskox. Only last minute efforts by individuals or governments have saved these animals.

The distribution of the plains bison before the North American Prairies were settled is shown in Figure 2.77, together with the major locations of present-day protected herds. There might be no bison today had not an Indian and a fur trader saved the few that remained.

In 1873 an Indian saved four calves from slaughter, and the next year a Winnipeg fur dealer saved another five calves. By 1914 the descendants of the four calves saved by the Indian numbered 745. Those saved by the fur trader numbered 87. These 87 offspring were released in the Wainwright Buffalo Park in Alberta, and by 1954 their numbers had increased to 40 000. This was a great victory for conservation in Canada.

Today there are about 28 **endangered species.** Any form of wildlife that is close to extinction is considered an endangered species. These endangered animals may not be hunted and, to encourage their reproduction, special breeding areas have been set up for some of them. Figure 2.79 shows some of these endangered species. In addition, there are 21 threatened and 44 rare species of plants and animals.

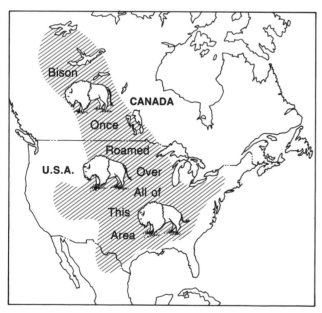

Figure 2-77 The distribution of the plains bison before the plains were settled

Figure 2-78 A bison

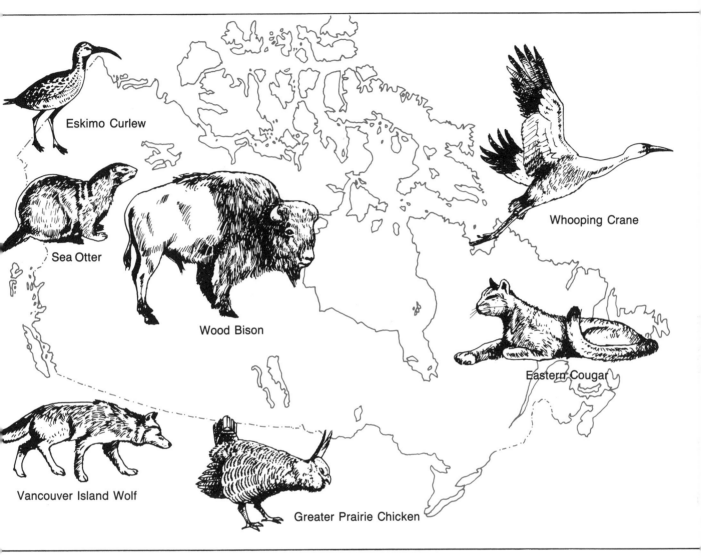

Eskimo Curlew

Whooping Crane

Sea Otter

Wood Bison

Eastern Cougar

Vancouver Island Wolf

Greater Prairie Chicken

Figure 2-79 Some endangered wildlife in Canada

58. The main causes of diminishing numbers of wild animals are listed below.
   (a) Choose one of these causes and explain fully the ways in which it leads to diminishing animal numbers.
   (b) Write down which of the causes listed below is (are) taking place near to your home. Explain where you have noticed any of these problems. What effects on wildlife, if any, have you personally observed?

   • Overhunting
   • Uncontrolled fishing
   • Clearing land for farms or cities
   • Draining marshland
   • Mining and smelting
   • Water pollution

National and Provincial Parks

Thirty-one National Parks and hundreds of Provincial Parks have been established across Canada. The total area contained in these parks is approximately 141 000 km², or more than 0.5 ha (hectares) for each resident of Canada. The parks include very different types of land, as you can observe from the photographs of some of the National Parks shown in Figure 2.80.

The main aims of these parks are to provide recreational and educational facilities for present and future generations. They will also ensure that some areas of the country are preserved in a natural state. In the 1985-86 period, the cost of staffing and maintaining the National Parks worked out to about $4 for each Canadian.

Figure 2-80 Location of Canada's National Parks

59. Study the photographs of the National Parks shown in Figure 2.81. Set up a table like the one below. Fill in the information using the physical regions map on page 3 (Figure 1.3).

NAME OF NATIONAL PARK	PHYSICAL REGION	VEGETATION REGION	DESCRIPTION OF PHOTOGRAPH
	SAMPLE	ONLY	

60. Do you consider that the National and Provincial Parks are a worthwhile investment of the Canadian taxpayers' money? Give reasons for your answer.

Figure 2-81 Canada's National Parks

(a) Pacific Rim

(b) Kluane

(c) St. Lawrence Islands

(d) Georgian Bay Islands

(e) Fundy

(f) Mt. Revelstoke

(g) Jasper

(h) Banff

Terra Nova National Park, Newfoundland

Terra Nova National Park is located on the east coast of Newfoundland. It consists of an area of irregular hills, some of them very steep. The shoreline is generally rocky and, in addition, there are many freshwater lakes. Vegetation types include dense forests, open rocky areas, and some marshland. There is a great variety of wildlife, including moose, many small mammals, birds, and various creatures like crabs and green sea urchins living along the seashore.

A map of part of the northern section of the park can be seen in Figure 2.82. There are many deep seawater inlets, called **sounds**, like this in the park. These sounds, or inlets, were carved by the ice that moved over the area. When the sea rose, they were flooded.

61. Figure 2.82 shows the northern section of the park which, as you can see, has been developed very little.
    (a) Trace the coastline, lakes, and rivers onto a blank piece of paper. Mark onto your map, the Trans-Canada Highway, existing roads, trails, and park facilities.
    (b) Imagine that you have been hired to make a plan for developing this area of the park. Bearing in mind the reasons why National Parks have been established (see page 78), you should carefully put your ideas onto the map. Use suitable symbols and a key of explanation. You might wish to include campgrounds, nature trails, boating and swimming facilities, canoe routes, and portages. At present, the most popular activities in the park include swimming, canoeing, boating, hiking, fishing, summer and winter camping, cross-country skiing, and snowshoeing. One park rule that may affect your design is that no powered boats are permitted on freshwater lakes and

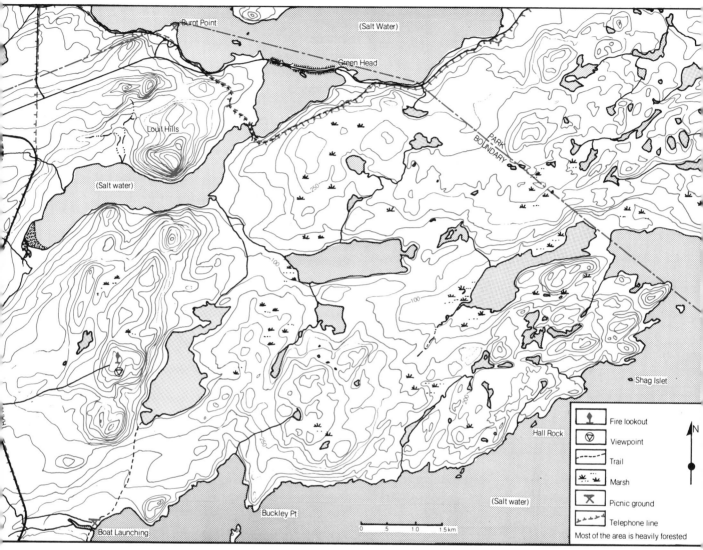

Figure 2-82 The northern part of Terra Nova National Park in Newfoundland

streams. Your design should include facilities that would be suitable for family groups, teenagers, young adults, and older people.

(c) In your notebook list and briefly describe the facilities that you have included in your plan. Indicate for which group(s) each facility would be suitable (family groups, teenagers, young adults, older people).

Vocabulary

Magma
Igneous rocks
Granite
Weathering
Erosion
Sedimentary rocks
Sandstone
Fossils
Metamorphic rocks
Gneiss
Glaciers
Scale
Line scale
Written scale
Ratio scale
Large scale maps
Small scale map
Orient
Cardinal points
Topographic maps

Symbols
Key or Legend
Military grid
  co-ordinates
Cartographers
Contour line
Contour interval
Weather
Atmosphere
Climate
Precipitation
Climograph
Maximum
Minimum
Range of temperature
Altitude
Water vapour
Evaporates
Condense
Orographic precipitation

Convectional precipitation
Cyclonic precipitation
Natural vegetation
Tundra vegetation
Tree line
Conifers
Deciduous tree
Irrigated
Mineral matter
Humus
Soil profile
Podzolic soils
Black soils
Species
Extinct
Habitat
Migrate
Hibernate
Endangered species
Sounds

Research Questions

1. Make as accurate a sketch map as you can of the area around your school or home. If you are in a built-up area, include details of roads, important buildings, shops, etc. If your map is or a rural area, include roads, buildings, and field boundaries. Work out an approximate scale with the help of your teacher, and include it on your map. Also include a title, a north sign, and a key (legend). Use suitable colours and symbols where possible.
2. Go to an area that has as "natural" a state as you can find.
   (a) Describe the area, referring to the shape of the land, the density of vegetation cover, and the uses to which the land is usually put.
   (b) Identify and list as many of the plants and trees as you can. State which types cover the greatest area.
   (c) If you can find a place where the soil is exposed, draw a carefully labelled soil profile.
   (d) List the animals that you see, together with signs of animals such as birds' nests, gopher holes, etc.
3. Using your school or local library for information, choose one Canadian animal to study. Then describe the environment in which it lives and explain how it is suited to its environment. Include information about its food, shelter, and habits.

4. Obtain information about one National or Provincial Park. Construct a display suitable for mounting on the wall. Your display should include the following:

(a) A large, neatly written title, using the name of the park and the province or territory in which is is located.

(b) Photographs, with a brief description of each, showing
   (i) scenery in the park
   (ii) special plants
   (iii) animals
   (iv) recreational facilities and activities
   (v) educational exhibits (if any exists).

(c) A map locating the park in Canada or in your province.

NOTE: If you write away for free pamphlets, ask for two of each. This is so that you can use the pictures and the information printed on both sides of the pamphlet.

3 OUR HUMAN HERITAGE

The First Canadians

Less than 400 years ago, the area we know as Canada lay quiet and almost empty. The human population consisted only of 220 000 native people.

These first inhabitants of Canada were the Indians and Inuit. Essentially, the Indians occupied the land south of the tree line, while the Inuit lived north of this line. The word "Inuit," it should be noted, means "the people." They prefer this name to "Eskimo," which means "eater of raw flesh."

The Indians and the Inuit led very different lives and very seldom mixed together. Some experts believe that these two groups are not related to each other at all. There is, nonetheless, one characteristic that the Indians and Inuit have in common, as you will discover.

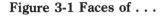

Figure 3-1 Faces of . . .

 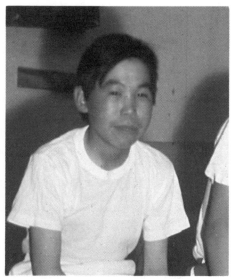 

(a) An Indian   (b) An Inuit   (c) A Person from Southeast Asia

---

1. (a) Look at the photographs in Figure 3.1. What facial features of the Indian and the Inuit are similar?
   (b) In what ways does the face of the Southeast Asian look like that of the Indian and Inuit?

There is one theory that clearly explains the relationship between these three groups. About 10 000 years ago there was an Ice Age. At that time a land bridge joined Asia and North America. Many Indians and Inuit wandered across this land bridge, probably following and hunting large herds of animals. Once they reached North America, the Inuit spread out across the Arctic and into Canada. In contrast, the Indians moved south into the forests of Alaska, to British Columbia, and then to the rest of southern Canada.

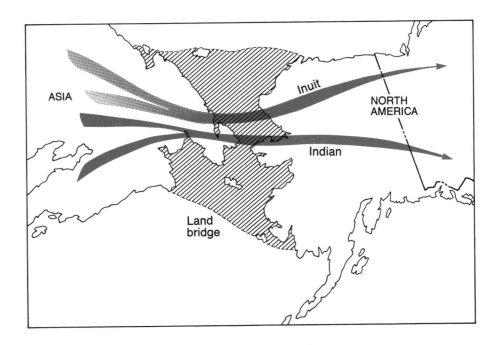

Figure 3-2 The land bridge connecting Asia and America during the Ice Age

ASIA

Inuit

NORTH AMERICA

Indian

Land bridge

2. (a) Using your atlas, locate the area where the land bridge is thought to have existed. What name is given to the narrow neck of water that exists there today?
    (b) How wide is that body of water? What problems would people have in trying to cross it today?
    (c) What impact would there be in North America today if there was a land bridge at that point? Refer to the map of the Arctic Ocean in your atlas.

Figure 3-3 Rocky land with snow patches all the way to the horizon
This scene is a typical Arctic landscape, similar to what you would see from the window of an airplane.

An Arctic Adventure

Imagine that you are flying in an airplane over the Arctic region of northern Canada. It is winter and as you look out of the plane window, all you see is rocky and stony land with snow patches all the way to the horizon. The sun is very low in the sky and there are a few snow clouds that the plane cuts through on its route.

The pilot then announces that the temperature outside is −20°C. He also tells you that the windspeed on the ground is 50 km/h. You are very glad that you are in a warm plane. All of a sudden, one of the engines catches fire and you see the flames leaping out from the wing just by your window. The plane begins to shake. The pilot calmly announces that he will attempt a crash landing on the ice below. Fasten your seat belts!

As the plane shudders and sinks lower and lower, you feel your stomach twist into knots as you think of the crash that is coming... The next thing you remember is waking up in the snow. You are lying on about 15 cm of snow, which is on top of about 60 cm of ice. The ocean lies beneath the ice. As you shake your head, you look around just in time to see the plane explode in a ball of fire. You cannot waste time, you must...

Complete the above story by filling in five steps you must take in order to survive in this frozen wilderness until help arrives. List these steps in order, starting with the most important one. Remember that absolutely nothing of the plane remains, except four or five small chunks of metal under 10 cm in length. There were 26 people on board. Of these, four are dead, two have serious head wounds, one has a broken leg, and twelve have minor cuts and bruising.

THE INUIT PEOPLE

The Inuit people have lived in the environment described above for many generations. Experience has taught them a great deal about survival.

The Inuit learned how to take advantage of every object in their environment. Figure 3.4 illustrates this fact. You will notice that animals were of much greater value to the Inuit than other parts of their environment.

Life for the 25 400 Inuit today is very different from that of their ancestors. Most of them live in houses similar to some found farther south in Canada. Modern communications bring to the Inuit knowledge of a world unknown to their grandparents. Inuit children, for example, are able to watch movies showing scenes from all parts of the world. Television can also bring news of Canadian and world affairs to many parts of the Arctic. The Inuit people may be better informed about southern Canada and the U.S.A. than many Canadians are about the Arctic.

Figure 3-4 How the Inuit people made use of all parts of their environment

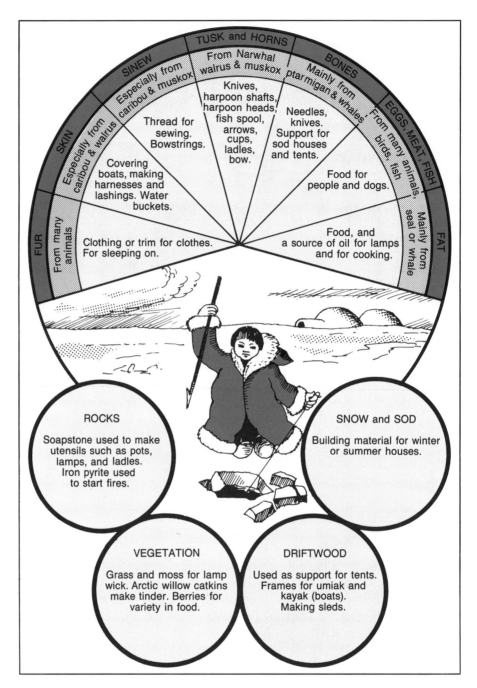

SINEW — Especially from caribou & muskox — Thread for sewing. Bowstrings.

TUSK and HORNS — From Narwhal walrus & muskox — Knives, harpoon shafts, harpoon heads, fish spool, arrows, cups, ladles, bow.

BONES — Mainly from ptarmigan & whales — Needles, knives. Support for sod houses and tents.

SKIN — Especially from caribou & walrus — Covering boats, making harnesses and lashings. Water buckets.

EGGS, MEAT, FISH — From many animals, birds, fish — Food for people and dogs.

FUR — From many animals — Clothing or trim for clothes. For sleeping on.

FAT — Mainly from seal or whale — Food, and a source of oil for lamps and for cooking.

ROCKS — Soapstone used to make utensils such as pots, lamps, and ladles. Iron pyrite used to start fires.

SNOW and SOD — Building material for winter or summer houses.

VEGETATION — Grass and moss for lamp wick. Arctic willow catkins make tinder. Berries for variety in food.

DRIFTWOOD — Used as support for tents. Frames for umiak and kayak (boats). Making sleds.

The arrival of modern technology and the white people's way of life in the Arctic has had a great impact. In the past, many problems resulted. Recently, the Inuit have used technology to keep important parts of their culture. For example, the Inuit Broadcasting Corporation is an advanced system set up by the Inuit people.

Northern communities, like many other ethnic groups, have problems when their culture meets other people's cultures. This is shown in Figure 3.5.

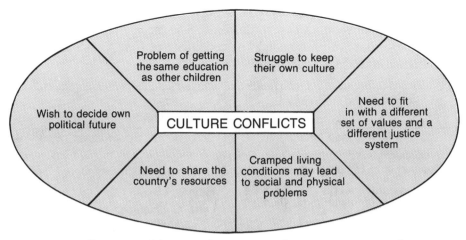

Figure 3-5 Some problems when one culture meets another culture

The Resolute Experiment and Its Background

One attempt to improve conditions at Resolute, on Cornwallis Island, was proposed in 1973. Resolute was made up of two separate communities. The Inuit community of about 200 people was situated about 6 km south of the "white" community, which was centred around the airport.

---

3. Find Resolute on an atlas map.
   (a) Describe its location in Canada.
   (b) Explain what effects its location would have upon the climate and the general way of life of the inhabitants throughout the year.
4. The most regular commercial air links are with Iqaluit (Frobisher Bay) and Montreal. What is the distance in kilometres from Resolute to each of these places?
5. The following steps will give you a clear idea of the isolation of communities such as Resolute.
   (a) (i) Obtain a base map of Canada, an atlas, a pair of compasses, and a pencil.
       (ii) Put the point of the compass on your home town. Draw a circle with the radius equal to the distance from Resolute to Iqaluit.
       (iii) Using your atlas, note the towns and cities the circle passes through or near to.
   (b) Complete the following sentence, using the results of 3 (a):

   If I lived in Resolute, the nearest town would be as far away as * is from where I live now.

   (c) Repeat the same exercise, this time using the distances between Resolute and the city of Montreal. Write your own sentence to explain your findings.

The chart below shows how the services for the Resolute residents were divided between its two settlements. The inconvenience of such a system, together with the poor living conditions of some residents, led to the decision to build a new town.

AT THE FORMER INUIT VILLAGE THERE WERE . . .	AT RESOLUTE AIRPORT THERE WERE. . .
200 permanent Inuit inhabitants. a nursing station, the nearest hospital is at Iqaluit (Frobisher Bay). an elementary school (to Grade 6); students of Grade 7 and up went south to school. a church.	250–600 non-permanent white inhabitants. a small Hudson Bay store. a movie theatre. an RCMP office. a bank.

The two settlements that existed at Resolute are shown in the following map and photographs (Figures 3.6 and 3.7).

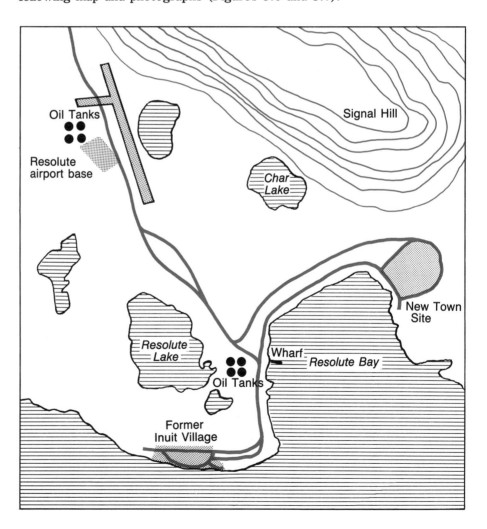

Figure 3-6 The location of the Resolute settlements

Figure 3-7 The former Inuit village and the airport base

6. If you had been brought up as a resident of Resolute, in what ways would your life have differed from what it is now? Your answer should include reference to
   (a) your daily needs of food, clothing, and shelter
   (b) your education
   (c) your leisure activities
   (d) health care and emergency treatment.

Isolation is a major problem in many northern communities. This isolation, and the severe climate, means that food, fuel, and other supplies have to be brought in annually from southern Canada by three types of ships, which are as follows:
(1) an icebreaker (which also serves as a hospital ship)
(2) a general supply ship bringing food, vehicles, and building materials
(3) a fuel tanker bringing heating, general, and aviation fuels

Figure 3-8 (a) Supplies being unloaded at Resolute from the supply ship

Figure 3-8 (b) Planes are also used to bring supplies when conditions are suitable

7. Supply ships to the eastern Arctic originate in Montreal.
   (a) Using your atlas to help you, carefully describe and mark on a base map of Canada the route that the supply ships would use to get to Resolute.
   (b) Explain why deliveries usually occur in August.
8. Some commodities are flown in from Montreal.
   (a) What items would be flown in, and why?
   (b) Why are supplies not flown in whenever they are needed?

The New Town

The site for the New Town is shown in Figure 3.6. The town was to provide housing for both the Inuit and white people in the area.

The particular advantages of the site are listed below.

- It is sheltered by Signal Hill from the cold north winds.
- It is away from the flight path for planes to and from the airport.
- It receives as much sunlight as possible.

A plan of the New Town site is shown in Figure 3.9.

Figure 3-9 Plan for the new town at Resolute

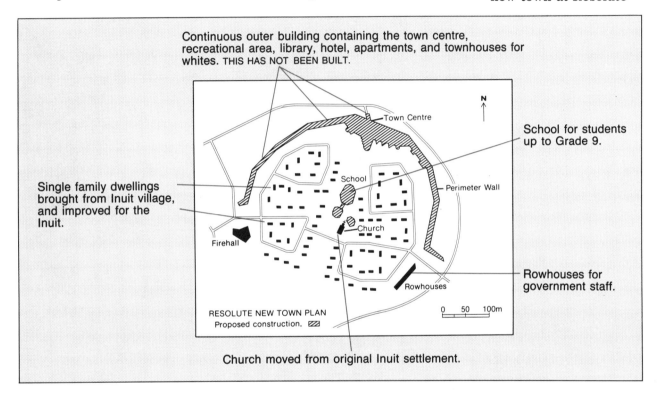

Continuous outer building containing the town centre, recreational area, library, hotel, apartments, and townhouses for whites. THIS HAS NOT BEEN BUILT.

Single family dwellings brought from Inuit village, and improved for the Inuit.

School for students up to Grade 9.

Rowhouses for government staff.

Church moved from original Inuit settlement.

The following progress table lists each year's activities.

PROGRESS FOR THE NEW TOWN					
1973	1974	1975	1976	1977	1978
Idea for new town proposed by the Northwest Territories (NWT) government.	Site was surveyed.	Site was prepared (sewers, hydro, etc. supplied).	Inuit houses moved from old village. Row houses built.	New fire hall built. Town centre designed.	Federal government withdrew funds for further development. NWT government paid for completion of water and sewage service.
PLANNING	WORK ON PROJECT				CANCELLATION

Figure 3-10 The partially completed new town of Resolute

Surveys and Mapping Branch © Her Majesty the Queen in Right of Canada with permission of Energy, Mines and Resources Canada.

9. (a) In Figure 3.10, locate the row houses and the church.
   (b) Comparing the photograph (Figure 3.10) and the map (Figure 3.9) on page 93, which direction is at the top of the photograph?
   (c) The water supply for the new settlement was to come from Char Lake, which you can locate in Figure 3.6 on page 90. Would the water-supply pipes lead into the town from the right or from the left in the photograph?
10. Explain what advantages and disadvantages would result from
    (a) combining the services in one community and making the school include students up to Grade 9, instead of Grade 6
    (b) including both Inuit and whites as members of the Town Council
    (c) having the white and Inuit people living in the same community.
11. The composition of the white community at Resolute is almost exclusively male, whereas the Inuit people have a balance of both sexes. Can you foresee problems which might arise as a result of their living in a compact community?
12. The federal government withdrew funds in 1978 because interest in oil exploration in the eastern Arctic had decreased. Do you consider that the federal government was justified in calling a halt to the project? Explain the reasons for your answer fully.

THE INDIAN PEOPLE

One theory suggests that after the Indian people arrived in Canada from Asia, they wandered over the land and settled in many different environments. Figure 3.11 shows the distribution of the Indian and the Inuit people in Canada.

The way of life that each group developed depended upon the availability of materials they could readily use. The Indians had to obtain all of their food, clothing, and shelter from their local environment. The Plains Indians, as well as other native groups, were inventive and resourceful in providing these necessities of life for themselves.

Figure 3-11 The distribution and environment of Indian and Inuit groups in Canada

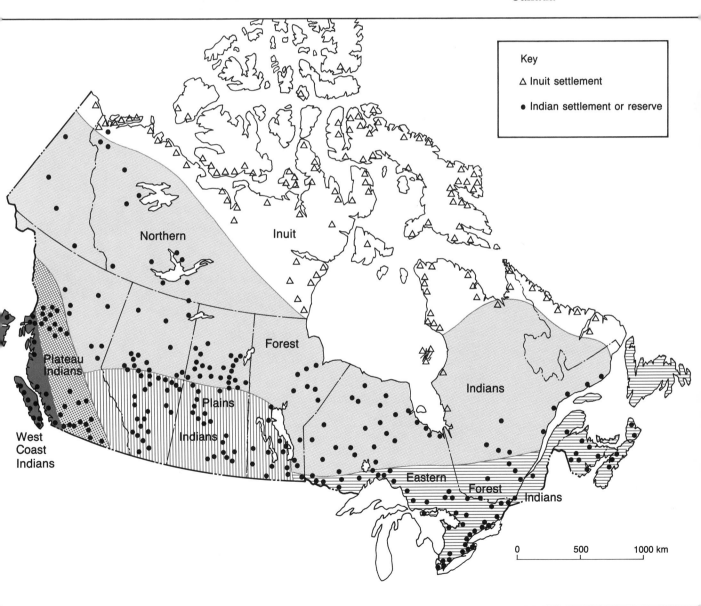

The Plains Indians

The Plains Indians inhabited what is now the southern Prairie provinces. Their environment consisted of rolling or flat land with few river valleys and little surface water. The climate was severe, with long cold winters and short hot summers. Much of the area was covered by grasses, with trees found only along streams and rivers. Animal life was abundant and included bison, antelope, deer, bear, many small birds, and small animals. Fish were not an important source of food.

Although the Indians hunted many types of animals, the bison was the most important. They employed many methods of hunting, from herding them into a pound, where they were killed with stone-headed clubs, to stampeding them over a cliff.

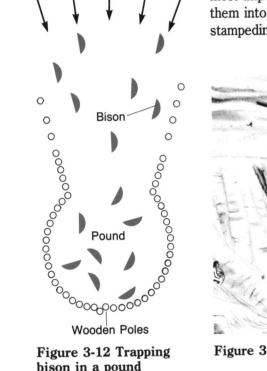

Figure 3-12 Trapping bison in a pound

Figure 3-13 Stampeding bison over a cliff

Sometimes a small number of hunters would cover themselves with bison or wolf pelts (skins). They would cautiously approach the bison until they were within range of the animals. At this point the Indians would carefully shoot their arrows to kill the selected bison. Other animals were also hunted using bows and arrows.

Later the Plains Indians used horses, which made hunting much easier. The meat from the hunt was either cooked for immediate use or sun-dried for later use.

Plant foods such as roots and berries supplemented (added to) the diet of the Plains Indians. Dried meat was sometimes pounded to a powder and mixed with pounded choke cherries to make **pemmican.** Pemmican would keep well and was easy to transport.

Tipis made of long poles and bison hides formed the basic lodge. The smoke from a central fireplace escaped from an adjustable opening at the top of the tipi.

Figure 3-14 The interior of Plains Indians' tipi

Figure 3-15 The horse as a beast of burden

Before horses were introduced into the area in the early 1700s, the family dogs were used to haul property. Horses later took over the job and enabled the Indians to transport many more items.

The Plains Indians used clothing made from the skins of various animals sewn together with strips of leather or sinews. They were often decorated with porcupine quills and teeth. The hides were scraped and rubbed to make them soft and supple (workable).

---

13. Using your school or local library, study one *other* major group of Indians, as shown in the map, page 95 (Figure 3.11).
    (a) Describe the environment. (Refer to the maps in Figure 2.36, page 51, and Figure 2.60, page 65.)
    (b) Explain how they used their environment to provide food, clothing, shelter, and transportation.
    (c) Include diagrams and sketches where possible.

Figure 3-16 Clothing of the Plains Indians

THE INDIAN PEOPLE TODAY

As settlers spread over the land during the last few centuries, approximately 200 000 Indians found their traditional way of life slowly changing. The area of land they could hunt in diminished as white farmers and ranchers increased in number.

In order to make the change from a traditional way of life easier, the Indians were given certain areas to live in called reserves. However, many of the reserves were not able to support the needs of their inhabitants. By 1901 the Indian population had dropped to 100 000 because of influenza, pneumonia, measles, and other diseases.

Early attempts to give aid to the Indians, in the form of welfare, did not solve their problems. By comparison with Canadians in general, the Indians remained poor. Many of the reasons for this poverty are outlined in Figure 3.17.

Figure 3-17 Factors that led to Indian poverty in the past

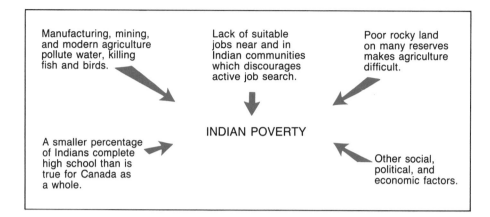

Manufacturing, mining, and modern agriculture pollute water, killing fish and birds.

Lack of suitable jobs near and in Indian communities which discourages active job search.

Poor rocky land on many reserves makes agriculture difficult.

INDIAN POVERTY

A smaller percentage of Indians complete high school than is true for Canada as a whole.

Other social, political, and economic factors.

The Canadian government and the Indian Band councils are attempting to improve conditions on the reserves and to provide better opportunities for employment.

Figure 3-18 Changes in access to toilet facilities and piped water on Indian reserves, 1977 to 1984 (percentage)

REGIONS	TOILET FACILITIES		PIPED WATER	
	1977	1984	1977	1984
Atlantic	83	96	84	93
Quebec	90	79*	91	95
Ontario	40	57	43	66
Manitoba	19	47	24	46
Saskatchewan	8	48	12	51
Alberta	53	66	53	60
British Columbia	89	97	97	98

*This drop reflects the difficulty of comparing data for the two years involved. There has probably been no significant change.

Figure 3-19 Houses on Indian reserves

Figure 3-20 A modern and a less modern Indian school

Recent estimates of Indian unemployment range from 35% to 75% of the labour force. The proportion of Indian children enrolled in elementary school is almost the same as the average for Canada as a whole. Enrollment of Indian students in secondary schools has more than doubled since 1965, but has been gradually declining since 1972-73 which was the peak year. In 1984-85, 66 percent of eligible young people were enrolled in Grade 9 and 31 percent in Grade 12. The government is attempting to provide better education and vocational (job-oriented) training.

High priority is also being given to treatment of alcohol and drug abuse. Much success has been achieved where community groups have attacked these problems. Improvement in employment and social conditions will probably result in the reduction of the frustrations that lead to social problems which have plagued the **reserves.**

Figure 3-21 Business ventures *Business ventures such as this gift shop provide income and employment for many native people.*

Today in Canada there are over 300 000 registered or status Indians. Altogether there are four main groups of native people.

STATUS INDIANS
are registered as
Indians under the
Indian Act.

NON-STATUS INDIANS
are people who are Indian by birth,
heritage, and culture but are not classified as Indian according to the Indian Act.

METIS
are people of
mixed Indian and
non-Indian ancestry.

INUIT
are the native people
who are mostly located
in the Northwest Terri-
tories, Labrador, north-
ern Manitoba and north-
ern Quebec. They have
a different culture from
the Indians.

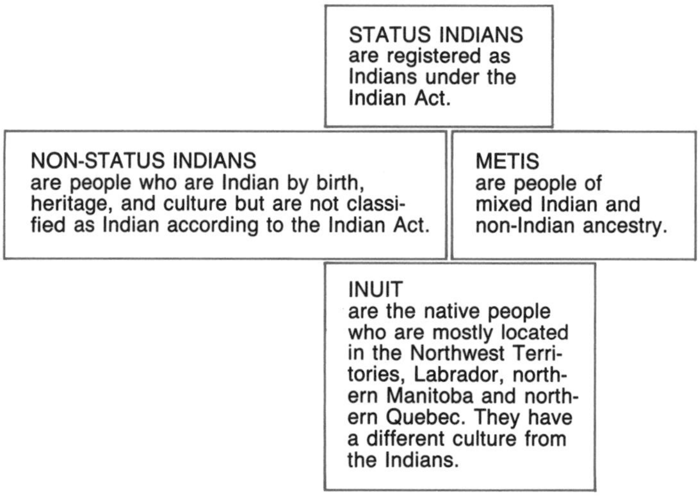

Figure 3-22 The four main groups of native people

The Indian and Inuit people of Canada were the first citizens of this land. They provide us with a unique cultural heritage. This means that they have developed special types of art, crafts, beliefs, and ceremonies. It is only just that they should have the same benefits and standards of living as all Canadians. Achieving this will involve continued efforts by the government, the Indians, and all Canadian citizens.

14. Give examples of the unique cultural heritage of the Inuit and Indian people.
15. Why do you think that the government considers it important for the Indian people to have control over improvements on their own reserves?
16. In what specific ways have the modern Inuit and Indians experienced the same problems?

More Recent Immigrants

Long after the Indians and Inuit settled in Canada, many immigrants began to arrive from western Europe. Many millions of people emigrated (left) from their home country to become immigrants (new arrivals) to Canada.

The only way for the early immigrants to reach Canada was by ship. Unfortunately, the journey by ship became a nightmare for some people. The conditions for the poorer passengers were so deplorable that many died before they reached Canada. Contaminated drinking water, rats, and filth caused many of these problems. When someone died, the body was quickly buried at sea to prevent the spread of disease.

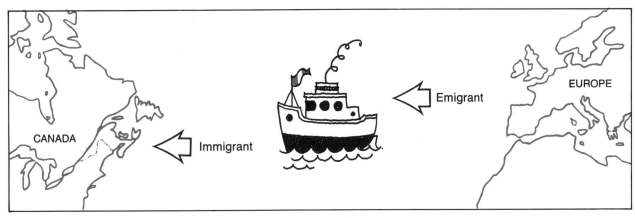

Figure 3-23 Emigration and immigration

    Upon arrival in Canada, many people were so sick that it took many weeks for them to recover. At that time, however, Canada could not offer adequate medical care and some immigrants never regained their good health.

    Why would anyone want to come to Canada? This question can be partially answered using the diagram below (Figure 3.24).

Figure 3-24 Reasons for immigration into Canada

17. Identify the push and pull factors for each of the following immigrants:
    (a) A young Polish man who left his homeland when it was taken over by the U.S.S.R. He chose to come to Canada because of its climate and its *multicultural society*. A multicultural society is one that has a large number of immigrant groups who still retain many customs from their homelands.
    (b) A West Indian student who came to Canada to get a university education and decided to stay because of a better chance of employment.
18. With the help of your relatives, construct a family tree. Include names, dates, employment, dates of emigration, etc. as shown in the following diagram.

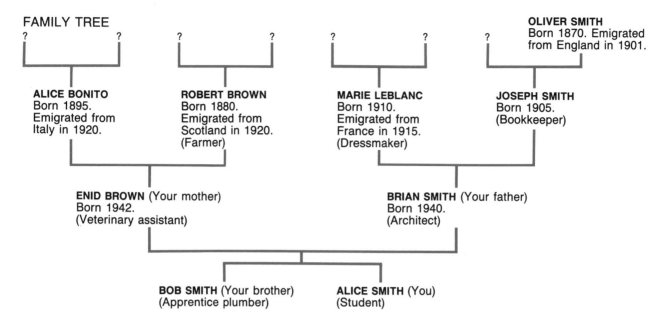

FAMILY TREE

?   ?   ?   ?   ?   ?   ?

OLIVER SMITH
Born 1870. Emigrated
from England in 1901.

ALICE BONITO
Born 1895.
Emigrated from
Italy in 1920.

ROBERT BROWN
Born 1880.
Emigrated from
Scotland in 1920.
(Farmer)

MARIE LEBLANC
Born 1910.
Emigrated from
France in 1915.
(Dressmaker)

JOSEPH SMITH
Born 1905.
(Bookkeeper)

ENID BROWN (Your mother)
Born 1942.
(Veterinary assistant)

BRIAN SMITH (Your father)
Born 1940.
(Architect)

BOB SMITH (Your brother)
(Apprentice plumber)

ALICE SMITH (You)
(Student)

Case Study: The Moon Family

Mr. Moon was born and raised in South Korea, just outside its capital city of Seoul. After Mr. Moon married, he moved to Seoul to teach music in a private school.

As a teacher he enjoyed a high standard of living relative to other Koreans. Mr. Moon lived in a middle-class neighbourhood in Seoul not far from his friends and relatives.

At that time his wife was also a teacher. This meant that the income for the Moon family was high enough to allow them to buy a number of luxuries unknown to most Koreans.

As a hobby Mr. Moon was involved in making xylophones. He also worked with music groups such as choirs in Seoul.

During the early 1970s, Mr. Moon and his wife became very concerned about their future in South Korea. They felt South Korea was too crowded for them, with little room to live and raise a family. By that time they had a family of three boys. Concern about the education of their sons strongly influenced the Moons' decision to leave South Korea. There was limited opportunity for advanced education in Seoul. In addition, Korean laws were considered to be very strict.

In 1974 the Moons finally decided to emigrate to Canada. Since an uncle of Mr. Moon lived in Mississauga, that is where the Moons decided to live.

In July 1974 when the Moon family stepped off the airplane in Toronto, they could speak almost no English. With the help of Mr. Moon's uncle and other Koreans adjustment to Canadian life began.

Learning the English language and adjusting to Canadian food were two main concerns for the Moons. Despite these adjustments, they kept many Korean traditions, such as respect for their elders and interests in

Figure 3-25 The Moon family

Korean art. Other traditions include celebrations of Korean holidays, speaking Korean at home, and practising Korean music.

Not long after arrival in Canada, Mr. Moon began to work in a paint factory in Toronto. After two years, however, he had saved enough money to buy his own dry cleaning store, where he now works. Mr. Moon works sixty-four hours per week in his store. He also pursues hobbies of music and carpentry.

Mr. Moon hopes that his family will have a bright future in their new home — Canada. His sacrifice in leaving South Korea has set out a new life in Mississauga for his family.

19. (a) What were the push factors in the move of the Moon family?
    (b) Describe the pull factor(s) that brought the Moons to Canada.
20. What sacrifices did Mr. Moon make in coming to Canada?
21. What attempt is Mr. Moon making to adapt to Canadian life?
22. In what way could Canada be considered to be richer as a result of the Moon family's arrival?

Immigrants coming to Canada bring with them a great variety of ethnic traditions.

Figure 3-26 Ethnic traditions are proudly displayed by immigrants to Canada

They enrich our country by bringing along their **cultural baggage.** Figure 3.27 will help you to understand the meaning of this expression.

Figure 3-27 Cultural baggage

The cultural baggage of immigrants varies according to their country of origin.

23. (a) **In your own words, explain what is meant by "cultural baggage."**
    (b) **What cultural baggage, if any, do you have in your own family?**
24. **Copy the following table into your notebook. Fill in the table using evidence you have seen near your home in cities, towns, businesses, or farms. Some of your friends may have cultural baggage in their families, which you could also include.**

ETHNIC GROUP (Italian, German, etc.)	CULTURAL BAGGAGE
1. 2. 3. 4. **SAMPLE**	**ONLY**

Some traditions that are brought to Canada do not last long. Children born in Canada to immigrant parents, for example, often do not learn the language of their parents.

There are, however, many traditions that stay within a family for many years after they come to Canada. Religion is one aspect of life that may remain with an immigrant family for many generations. In Cape Breton Island, Nova Scotia, many people still attend the Presbyterian church, which was the church of their Scottish ancestors.

The map of Saskatchewan shown in Figure 3.28 illustrates the tremendous variety of immigrant groups that have populated this part of Canada.

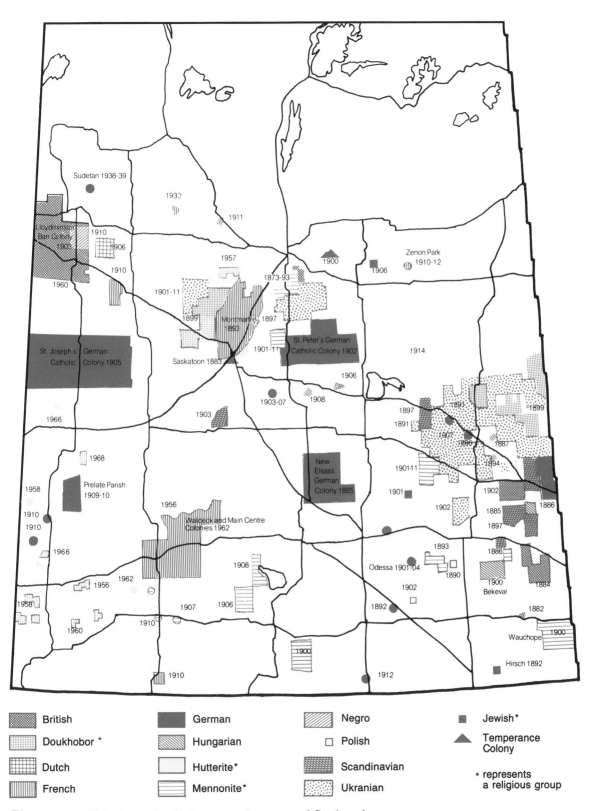

Figure 3-28 Ethnic and religious settlements of Saskatchewan

British	German	Negro	Jewish*
Doukhobor *	Hungarian	Polish	Temperance Colony
Dutch	Hutterite*	Scandinavian	* represents a religious group
French	Mennonite*	Ukranian	

Between 1885 and 1910, approximately 450 000 immigrants came to Saskatchewan, almost all of them to farm the promising wheat lands. Many religious groups also took the opportunity to move away from their home countries. They did this so that they could worship and live in the way that they believed was right.

---

25. (a) Using the map of Saskatchewan (Figure 3.28), list the four groups that occupy the largest blocks of land.
   (b) List four groups that settled in smaller areas of land.
26. Explain why settlement was almost all in the southern part of the province.
27. Explain in what ways the arrival of these immigrants would have affected the lives of the Plains Indians.

The number of people who come to Canada varies from one year to the next. It is the strength of the push and pull factors that determines how many people come each year. Also, Canadian immigration officials apply tighter rules to ensure that today's immigrants have better qualifications than ever before.

Figure 3-29 Average annual numbers of immigrants and the factors that affected their decision to come to Canada

---

28. Using the information from Figure 3.29, write a half-page account describing the variations in immigration between 1926 and 1984. Include information about the reasons for these variations.

YEARS	AVERAGE NUMBER OF IMMIGRANTS PER YEAR	FACTORS THAT INFLUENCED IMMIGRANT NUMBERS
1926-1930	146 290	Economy of Canada healthy. Many jobs available.
1931-1938	16 280	Depression.
1939-1945	12 750	Second World War.
1946-1950	86 078	Many people in Europe very poor and unable to raise fare money.
1951-1960	157 484	Economy of Canada very healthy. Many jobs available. Many refugees from Hungary accepted.*
1961-1963	79 809	Short recession in Canada.
1964-1975	167 054	Economic growth in Canada. Demand for labour. Many refugees from Czechoslovakia accepted.*
1976-1977	132 172	Poorer economic conditions in Canada. Stricter government restrictions.
1978-1984	113 258	Poor economic conditions and strict restrictions continue.

*Since World War II many people have come to Canada to escape unfavourable political conditions in their own countries. In these cases, the push factors were much stronger than the pull factors attracting people to Canada.

Not only have the numbers of immigrants varied, but their places of origin have also changed.

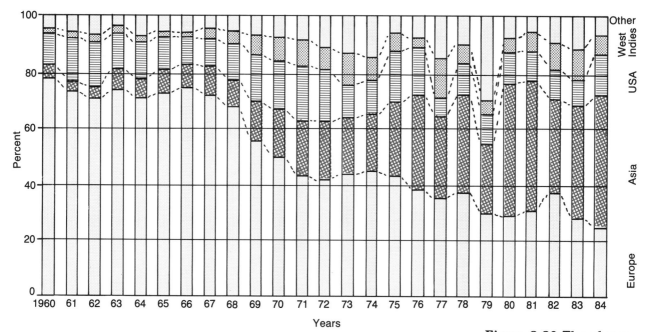

Figure 3-30 The places of origin of Canadian immigrants (1960-1984)

29. Using Figure 3.29 as a guide, describe the trends (changes) that you can observe in immigration from
    (a) European countries
    (b) Asian countries.

30. The figures below are the numbers of immigrants received by each part of Canada in 1984.

Newfoundland	297	Manitoba	3 906
Prince Edward Island	108	Saskatchewan	2 150
Nova Scotia	1 035	Alberta	10 665
New Brunswick	604	British Columbia	13 180
Quebec	11 598	Yukon and	
Ontario	41 518	Northwest Territories	116

(a) Using a blank map of Canada, devise a method of showing the destination of immigrants in 1984.
(b) Describe and attempt to explain
    (i) why most immigrants go to Ontario, Quebec, or British Columbia
    (ii) why Alberta receives most of the immigrants who go to the Prairies.
    Refer to the natural resources and manufacturing maps in your atlas as a guide.

107

Each major settlement in Canada has its own distinctive composition of people. Once a large number of a particular group become established in a city, they attract others from their home country. This gives the new immigrant a greater sense of belonging and security, as with the Moon family.

CITY	BRITISH	FRENCH	CHINESE	ITALIAN	GERMAN	OTHER
St. John's	93	1	<1	<1	<1	5
Charlottetown	81	6	<1	<1	<1	12
Halifax	71	7	<1	<1	3	18
St. John	75	11	<1	<1	1	12
Quebec	3	94	<1	<1	<1	2
Toronto	46	3	3	10	3	35
Winnipeg	36	8	<1	2	9	45
Regina	39	4	1	1	18	37
Calgary	49	4	3	2	8	34
Vancouver	49	3	7	2	6	33
<1 means less than one						

Figure 3-31 The percentages of selected immigrant groups in some Canadian cities

31. (a) Mark each of the cities listed in Figure 3.30 on a blank map of Canada.
    (b) Plot a small bar graph using the data in Figure 3.30 for each of the cities to show their ethnic compositions. A sample graph is given below.
    Fit the graphs onto the map using a line to connect each graph to the city it represents.

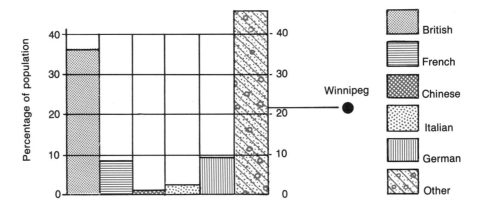

32. (a) Group the cities according to the following criteria:
    (i)   Those that have more than 75% of English origin.
    (ii) Those that have more than 75% of French origin.
    (iii) The city with the largest Italian community.
    (iv) The four cities with the highest percentage of German immigrants.

(v) The city that has the greatest percentage of Chinese immigrants.

(b) Use about half a page to describe the distributions of the selected immigrant groups in cities across Canada. Use your information from 32 (a) to help you with this.

(c) What other immigrant groups would probably make up the "Other" group in Winnipeg, Regina, and Calgary? The map of Saskatchewan (Figure 3.28) should help you to answer this question.

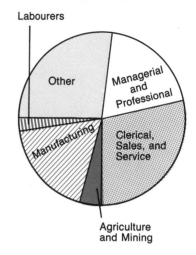

Figure 3-32 A pie graph showing the occupations of working immigrants

The guidelines we use in choosing who to allow into the country as immigrants are of great benefit to Canada. Figure 3.32 shows the types of workers who came into Canada in 1984.

Notice that over half of the immigrants are professional or skilled workers. Many of them had received years of advanced education and training before coming to Canada. As a result of this, our country benefits from their training without having to pay for it.

A few people emigrate from Canada. Some of them return to their original country. Some others move to places where their salaries will be higher. In particular, some professional people, such as doctors, move to the United States.

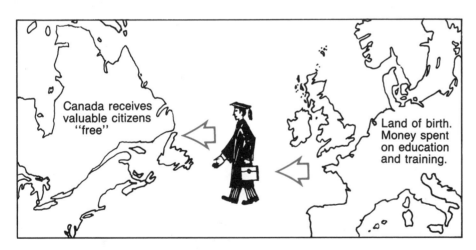

33. Why might people leave Canada to return to the country from which they emigrated?

34. (a) In the front of your atlas you will find a list of the various countries of the world. Listed as well is the per capita Gross National Product. The lower the number, the poorer the country is. List ten countries that would suffer the most from having trained people emigrate to Canada.

(b) Should Canada accept immigrants who have been trained in these poor countries? Explain your answer.

If a person wants to live in Canada on a permanent basis, it is necessary to apply to become a landed immigrant. A landed immigrant is a person who wishes to live in Canada and who is judged likely to become a valuable Canadian citizen. Immigrants who have lived in Canada for three years may apply for Canadian citizenship. To become a Canadian citizen with the same rights and responsibilities as all Canadians an immigrant must

(1) speak either English or French
(2) have a knowledge of Canada
(3) know what is expected of a citizen
(4) know the benefits of being a citizen
(5) take the oath of citizenship.

The people shown in Figure 3.33 have just made their oath of citizenship. It is a very important day in their lives, ensuring them of an equal place in a free society.

Figure 3-33 A citizenship court *This citizenship court was held in a school. From left to right, you see local dignitaries, the judge, the candidate, and an RCMP officer.*

Pemmican
Reserves
Emigrated

Immigrants
Multicultural society
Cultural baggage

Landed immigrant
Canadian citizenship

Research Questions

1. Write an illustrated account of one of the following:
   (a) The hunting methods used by native people in Canada.
   (b) Present day arts and crafts of our native people.
   (c) Methods of transportation used by the Indians and Inuit.
   (d) The construction of shelters by the Indians and Inuit.
2. Using information from your school or local library, describe the life of some early immigrants.
3. Interview a recent immigrant to Canada. Report the reasons why he or she came to Canada. In the case of a dependent, such as a child, it may be necessary to find out why the parents chose to come to Canada.
4. Investigate the history of a community near to where you live. Find out about the first immigrants, the countries from which they came, when they came, and their way of life.
5. Write a report of the history and present way of life of the Hutterites or Mennonites who migrated to Canada.
6. Using Figure 3.29 and the accounts of the happenings in Asia over the last ten years, explain the changing push and pull factors that have caused the increased numbers of Asian immigrants.
7. Find out the criteria (rules) used to determine whether a person will be allowed to immigrate into Canada. Study the rules and state which ones you would choose to change, giving your reasons.

4 WHERE WE LIVE

The Distribution of Canada's Population

There is so much land available in Canada that if all Canadians could be spread out evenly you could barely see your neighbour. The majority of Canadians, however, live in areas where the population is relatively concentrated. Figure 4.1 shows clearly where the centres of Canada's population are. As you will observe, the greatest concentration is along the shores of the Great Lakes and the St. Lawrence River in Ontario and Quebec. Other areas of fairly high concentration are found in the Prairies and along the coast and inland valleys of British Columbia and the Maritime Provinces.

Apart from these areas the population of Canada is sparsely distributed across the more northern portion of our country. A small number of centres are dotted across this northern area.

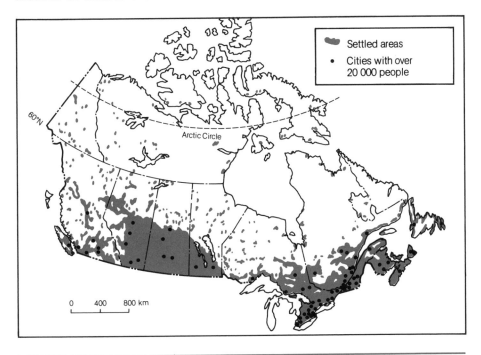

Figure 4-1 The distribution of Canada's population

1. (a) Examine Figure 4.1 and various *thematic maps* of Canada in your atlas. A thematic map is one that shows details about one topic, for example, the soils of an area. Suggest three reasons why Canadians are concentrated in urban areas of the country. Thematic maps showing climate and agriculture will help you.
   (b) Most Canadians live within 200 km of the United States border. In addition, the largest cities of Canada are found in this area. How does the location of Canada's population influence what Canadians watch on television and listen to on radio?

2. List the areas of Canada where few people live. Considering your answer to 1 (a), explain why more people are not attracted to these areas.
3. Across central and northern Canada there are a number of cities and settled areas surrounded by areas of sparse population. Refer to Figure 4.1 and the thematic maps of Canada in your atlas to explain why these areas of greater population density exist. Limit your answer to four specific areas or cities of Canada.

Land Survey Systems

Wherever people have settled in Canada, patterns of roads and fields have been established on the land. These patterns vary from one region to another. The physical geography of any region also has a significant influence on the settlement pattern.

QUEBEC: THE LONG LOT SYSTEM

The province of Quebec has a settlement pattern that is almost unique in North America. Just like most activities in Quebec, this pattern is centred on the St. Lawrence River. From the time of the first settlers in Quebec, the farms and human settlements have hugged both shores of the St. Lawrence.

4. Give two reasons to explain why the early Quebec settlers preferred to live next to the St. Lawrence River.

To understand how the people of Quebec settled on the land, imagine that you are transported back in time over 200 years. At that time it is your job to assign land to French settlers who have come to live in that area.

The land that you have control over is called a **seigneury** and you are called a **seigneur.** As a seigneur, you must divide your seigneury so that each settler has a suitable piece of land for farming.

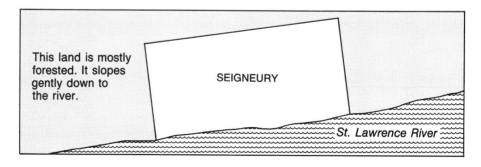

Figure 4-2 The seigneury that is to be subdivided

This land is mostly forested. It slopes gently down to the river.

SEIGNEURY

St. Lawrence River

5. (a) Copy the sketch of the seigneury into your notes. Use a full page.

   (b) Ten families arrive on your seigneury and each one wants some land. Divide the seigneury fairly among the ten families, remembering there are no roads, only forest. *Keep in mind the value of the river.* Be sure to number the lots 1 to 10, one for each family farm.

   (c) Mark a dot where you think each farmer would build a farmhouse so that it can be conveniently located.

   (d) Each *roture* (farm) was about 200 m wide and 2 000 m in length. On a piece of paper draw the boundaries of this roture using a scale of 1 cm represents 100 m. Draw in the St. Lawrence River at one of the narrow ends of the farm property. Read this description and make a map of the farm using colour and symbols.

   The farmhouse was set about 50 m from the river. It was 10 m by 15 m. There was a barn, which was about twice as big, close to the house. The two long sides of the farm were fenced and an extra fence had been built to provide a 5 m wide pathway along the edge of the property. This pathway allowed the farm animals to be led from the barn to the pasture without straying into the fields of grain.

   Near to the river, where there was the best soil, the farmer had a field in which he grew vegetables.

   Behind his farm, he had a field of wheat, and some fields for pasture and hay for winter storage. The back third of the property was left in forest and provided the farmer with firewood and timber.

   Include a line scale, a key, and a title with your map.

From the work you have done you can see how the landscape of Quebec was formed. The basic pattern of the land has essentially not changed since the first few years of settlement.

6. (a) When the land along the St. Lawrence River became crowded where would new farms be set out? Why?

   (b) What would have to be built to allow people to travel to and from the farms of this second *rang* (row) of farms?

   (c) Draw a sketch map which includes the river, three rangs, the farmhouses, and the roads.

7. (a) Put a piece of tracing paper over the air photograph (Figure 4.3).
   Mark onto your photograph: river (blue), roads (red), forest (green), fields (white), and buildings (black).
   The scale of the photograph is 1 cm represents 800 m.
   Include a scale, key, and title with your map.

(b) Imagine that you are on a boat travelling along that river. Describe the scenery which you would observe along the north shore.

Figure 4-3 The landscape of Quebec showing the field patterns

ONTARIO

The land survey system in Ontario is more complex than that in Quebec. Surveying is the accurate measurement of the land surface. In Ontario, the basic survey unit is the township, which is usually rectangular in shape. For each township a straight base line was measured.

Using the base line as a starting point for surveying, the township is divided into strips of land called concessions. Each concession has a road or concession line for access. To complete the transportation pattern side roads are built at right angles to the concession lines.

Each concession in a township is divided into lots; a farmer might own a whole lot, or part of a lot. Any farm can be located and given a concession and lot number in a township.

8. (a) Use a sheet of tracing paper to copy the various fields, roads, and farm buildings shown in Figure 4.5. Hold the tracing paper in place with paper clips.
   (b) Give the map you have drawn an appropriate title, a north arrow, and proper labelling of its features as listed in 8 (a).
9. Does the shape of the lots in Ontario have any advantages over the Quebec system? Explain your decision.

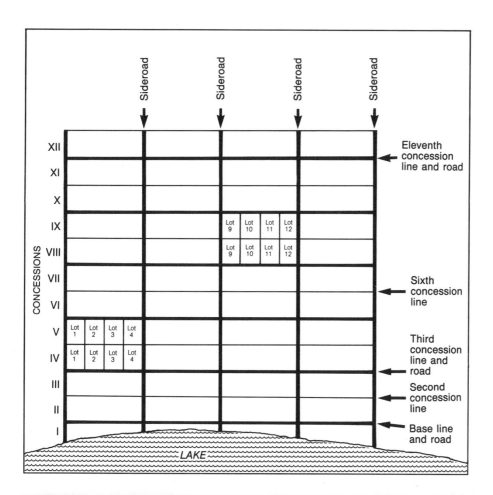

Figure 4-4 The system of surveying in an Ontario township

Figure 4-5 Air photograph showing some of Ontario's landscape

THE PRAIRIES

As you saw earlier, the Prairies are generally quite flat compared to the rest of Canada. This means that dividing up the land was relatively easy.

---

10. Draw a large square on a piece of blank paper. Assume that there are 16 settlers and that each one wants to have some land to farm. Divide up the land shown on your paper so each has a fair amount. Be sure to have enough roads so that all farmers can send out their crops easily. Also, use a small square to show each farmhouse.

The pattern of your map should look very similar to the air photograph in Figure 4.6.

Each one of the squares on your map is called a **section,** which is 256 hectares. A total of 36 sections would make up a township on the Prairies. It should be noted that not every farmer has a full section to work. On the average, however, Prairie farms are large compared to others in Canada.

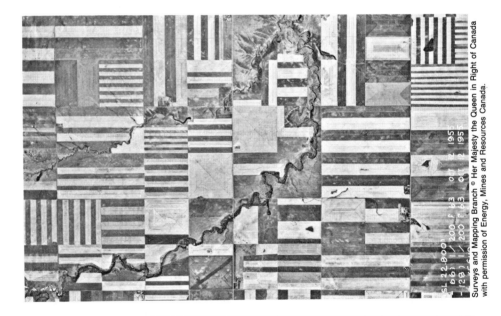

Figure 4-6 The Prairie survey pattern shows up clearly from the air

---

11. (a) Refer to Figure 4.6. What natural features disrupt the Prairie settlement pattern?
    (b) What problems do these features present for the farmers?
12. The air photograph (Figure 4.6) was taken in October after the wheat has been harvested. The stubble (base of the stems) in the fields is light coloured. What caused the black stripes?

Each survey system that we have looked at has its own unique qualities, but has its own problems as well.

13. Review the maps and air photographs of the different survey systems and then complete this table in your notes.

REGION OF CANADA	QUEBEC	ONTARIO	PRAIRIES
Distance of farmhouse from neighbours			
A square-shaped farm is most efficient to work. How efficient is each of these farms?	**SAMPLE**	**ONLY**	
How easy is the survey system to set up, relative to the others?			

14. Examine these three settlement patterns. Which of these systems do you feel suits a farmer's need to have an efficient farm, yet be close to his neighbours?

The Variety of Communities across Canada

When the first pioneers began to spread across Canada, they soon set up farms to grow crops for their own needs. In fact, these early farmers worked most of the land presently settled in Canada. Many **farmsteads** appear today much as they did years ago.

Each farmstead provided for a family such basic services as shelter, water, and food storage. During the early years of settlement in Canada, farmsteads were considered an important part of our economy.

Figure 4-7 A farmstead

15. Use a piece of tracing paper to draw a diagram of the farmstead shown in Figure 4.7. Label each part of the farmstead and outline the function (use) of each.

HAMLETS

Clearly, the farmstead could not provide all of the services and goods that a family required. As a result, small centres developed in the countryside for the benefit of surrounding farmers. Such a centre was called a hamlet. Hamlets may still be seen in most parts of southern Canada, although the functions of some of the buildings may have changed.

Even though hamlets have changed since early pioneer times, some still provide a basic level of service to the people that live nearby. Frequently, a hamlet may be called a Four Corners, although there may not be buildings on all corners of the settlement. A few of these hamlets have been abandoned, while others have barely changed. A small number have grown in size as they have sold more goods and added more buildings.

VILLAGES

A few of these small hamlets grew in size as they sold more goods and added more buildings. Gradually, such hamlets were transformed into villages. A village is larger than a hamlet, as you can see from the map of Palermo. Across Canada there are hundreds of villages, each serving its surrounding farms and residents. There could be 200 people living in a typical Canadian village.

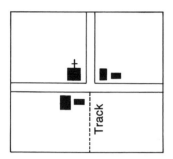

Figure 4-8 Maskawata Hamlet, Saskatchewan

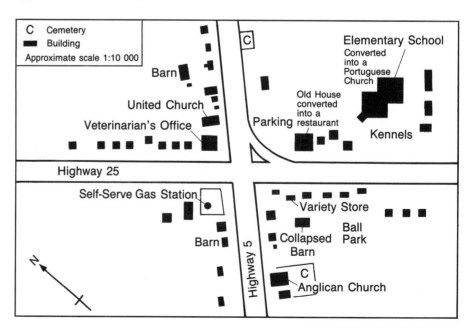

Figure 4-9 The village of Palermo, Ontario

120

Figure 4-10 Buildings in Palermo

16. (a) What specific changes occur in a hamlet to enable it to become a village?
    (b) What services would you find in a village that you would not find in a hamlet?
17. Examine carefully the photographs and map of Palermo. Using cardinal directions, describe the sector of the town in which each photograph was taken.

TOWNS

If a village continued to grow other houses would be added, as well as additional stores and services. A centre that is somewhat larger than a village is called a **town.** Major cities of Canada, such as Montreal, Halifax, Calgary, and Vancouver, were once towns. There are some centres today that still have the word "town" in their name. Gagetown, New Brunswick, Glovertown, Newfoundland, and Charlottetown, Prince Edward Island are good examples. A number of these centres have grown larger than towns in recent years.

Figure 4-11
(a) Edmonton, 1904,
when it was a town

(b) Edmonton today is a large and modern city

Road maps supply much information about the types of settlements in an area and the routes between them. As you can see from the map of northern New Brunswick, road maps also supply such additional information as tourist attractions, the quality of highways, services such as hospitals, and the sizes of settlements.

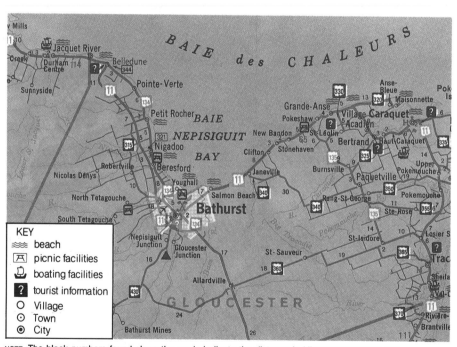

Figure 4-12 A portion of a road map of northern New Brunswick

KEY
≋ beach
🅰 picnic facilities
⚓ boating facilities
? tourist information
○ Village
◉ Town
◉ City

NOTE: The black numbers found along the roads indicate the distance in kilometres between two settlements.

18. What specific tourist attractions are displayed on the map (Figure 4.12)?
19. How many settlements on the map fall into each of these categories: hamlet, village, town?

20. (a) How far in kilometres is it from Belledune to Grande-Anse by road?
    (b) What highways would you use for this route?
    (c) What section of this route would be the slowest to drive? How do you know this?

GHOST TOWNS

A few towns and villages fail to survive and are abandoned. Just as a business can go bankrupt, so a settlement can lose its **economic base** and eventually become a **ghost town.** The economic base of a settlement includes all of the activities, like shops and factories, that bring money into the settlement.

Barkerville, British Columbia, is one example of such a ghost town. In 1863 there were over 10 000 people who lived in Barkerville, a centre built on gold mining in a forested wilderness. Unfortunately, the gold soon ran out and the miners left their houses, stores, and churches to return to their original homes. Furniture, wagons, clothing, and all types of other belongings were left behind. For years Barkerville lay empty with only rats and wild animals living in the buildings. Finally, in the mid 1960s the British Columbia government began to restore it to its original form.

Figure 4-13 Barkerville after restoration

21. (a) Explain why the sidewalks of Barkerville are raised. What improvements have been made in modern cities to avoid the need for these elevated sidewalks?
    (b) Why is the choice of wood for the buildings a logical one?

CITIES

A centre with a population of at least 10 000 is usually called a city.
Most Canadians live in cities of various sizes. Each city provides services and stores for a large region around it, as well as for its own residents. The number of Canadians who live in cities is increasing very quickly. This increase is due to the cities' own natural population increase as well as the movement of people from villages, towns, and other settlements.

22. (a) Examine the line graph shown below and then list the percentage of Canadians living in centres of 1 000 or more for the years 1871, 1901, 1931, 1951, 1961, 1971, and 1981.
    (b) During which period did the population of Canada's urban centres increase most quickly? How is this shown by the line on the graph?
23. Use the following statistics to construct your own line graph based on the design of Figure 4.14.

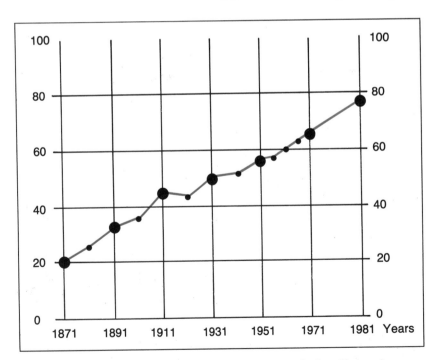

Figure 4-14 The proportion of Canada's population living in urban centres of above one thousand people

YEAR	PERCENTAGE
1871	3.1
1881	3.3
1891	8.2
1901	8.9
1911	15.0
1921	18.9
1931	22.4
1941	23.0
1951	23.3
1961	22.8
1971	26.7
1981	30.1

Figure 4-15 The percentage of Canada's population living in urban centres of above one hundred thousand people

24. (a) Compare the two graphs of Questions 22 (a) and 23. Which graph shows the greatest change between 1941 and the present day? What meaning does this have for Canada?

    (b) Looking once again at the figures for Question 23, how has the change shown here affected the way Canadians shop, go for entertainment, work, and spend their spare time?

CITY	POPULATION 1985	CITY	POPULATION 1985
Toronto	3 202 400	Halifax	290 600
Montreal	2 878 200	Windsor	249 800
Vancouver	1 348 600	Victoria	245 100
Ottawa-Hull	769 900	Regina	174 800
Edmonton	683 600	Oshawa	172 800
Calgary	625 600	Saskatoon	170 100
Winnipeg	612 100	St. John's, Newfoundland	160 700
Quebec City	593 500	Sudbury	147 600
Hamilton	559 700	Chicoutimi-Jonquière	139 400
St. Catharines—Niagara Falls	309 400	Thunder Bay	123 500
Kitchener	303 400	Saint John, New Brunswick	116 800
London	292 700	Trois Rivières	114 300

Figure 4-16 Canadian cities having over 100 000 people

NOTE: These figures are for the whole metropolitan areas of each city.

25. (a) In what province are most of these large cities located?

    (b) Which province has no city with a population of over 100 000? Explain the reasons for this.

26. (a) Calculate the percentage of Canadians living in the ten largest cities. To do this, add up the populations of the ten largest cities (to the nearest thousand). Then carry out this calculation:

$$\frac{\text{sum of population in ten largest cities}}{25\ 444\ 900\ \text{(population of Canada)}} \times 100$$

    (b) What does the answer to 26 (a) show us about the way of life of many Canadians?

27. (a) On a blank map of Canada mark and label the cities listed in Figure 4.16.

    (b) Join each city to the ones next to it using straight lines to represent highways. Compare this pattern of roads to that found in your atlas. What differences can you detect between the two? Give reasons for these differences where possible.

28. The six pictures in Figure 4.17 give some clues about the attractions of large cities. Write in essay form one half page to explain why cities grow.

29. (a) If you live in a city now, what reasons can you suggest for wanting to continue to live in the city?

    (b) If you do not live in a city at present, would you move into a city to live in the future? Give reasons for your answer.

Figure 4-17 The attractions of large cities

The Location of Cities

Canadian cities may appear at first glance to be located almost at random. There are, however, some important factors that determine where cities will flourish.

---

30. (a) Turn to a map of British Columbia in your atlas. List ten settlements whose names have some relationship to water. (Example: Creek, Port, Lake.)
    (b) Why is water so important for human settlement in British Columbia and in other places?
    (c) Referring to a map of Canada in your atlas, select five Canadian cities. Describe the location of each one relative to a named body of water.

One of the cities you may have chosen in answer to 30 (a) is Port Moody, near Vancouver, B.C. Its name suggests it has an important role in transportation. Such a port can be referred to as a **break-of-bulk** point, which means that goods are transferred from one mode of transportation to another. Wheat, for example, could be shipped to a break-of-bulk point by train and then transferred to a ship via a grain elevator. The entire Vancouver Port area is vital to the overseas shipping of Canadian wheat from the Prairies, especially to China and the Orient.

The land behind Port Moody, through which goods are shipped, is called the **hinterland.**

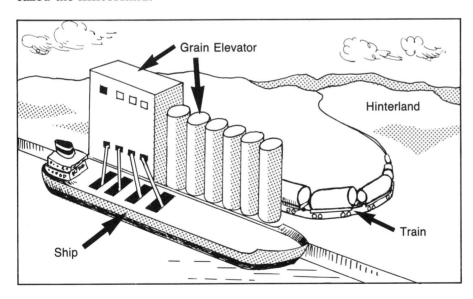

Figure 4-18 A break-of-bulk point

---

31. Refer to the base map (Figure 4.19) and the Canadian thematic maps of minerals and manufacturing in your atlas. Suggest the reasons for the location of those cities in Manitoba that are marked on the map.

Figure 4-19 Selected urban centres and railway lines in Manitoba

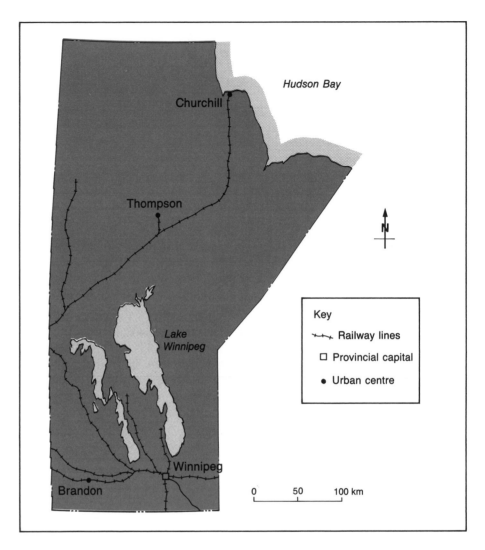

The location of Canadian cities can be summarized in the following way.

(1) Urban areas tend to be located next to bodies of water.
(2) Essentially, urban centres of all sizes, from hamlets to cities, are service centres for the surrounding land.
(3) Some centres have special functions, such as mining. (Example: Sudbury.)
(4) A number of large Canadian cities serve as ports or break-of-bulk points. (Example: Thunder Bay, Halifax.)
(5) For any centre to be founded, someone had to make a decision. An individual might have decided to set up a sawmill on a river, for example. That first decision could lead to the growth of a city.

The location of a city can also be examined from the point of view of its **site** and **situation.** The site refers to the local landscape, bodies of water, and surrounding land uses near where a city is built. A city's situation relates to its location relative to other cities, provinces, and countries, as well as transportation routes.

Figure 4-20 Vancouver, British Columbia

32. (a) Describe the site of Vancouver by referring to Figure 4.20.
    (b) What is Vancouver's situation? Describe its location relative to Canada, the U.S.A., other Canadian cities, and transportation routes. Use your atlas as a guide here.
33. Imagine that you are a town planner. You have been hired by the government to decide on the best location for a city. After some research, the choice of sites has been narrowed to five, as illustrated in Figure 4.21. Choose the site you consider to be the most favourable and supply reasons to support your decision. Explain why you rejected the other sites. In your research you should consider factors such as transportation, possible industry, water supply, and roughness of landscape.

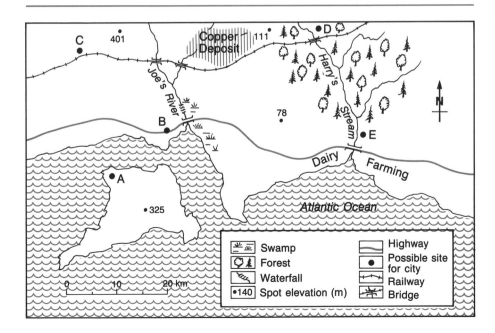

Figure 4-21 Possible sites for city development

129

Rural-Urban Fringe

Many Canadian cities are literally bursting at the seams with new arrivals. As land in the city itself is used up, new development spills out onto the countryside nearby.

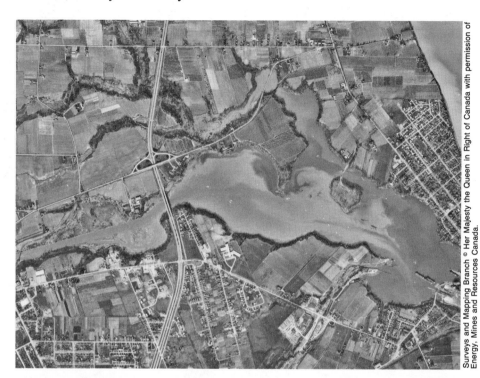

Figure 4-22 (a) St. Catharines, 1960

Figure 4-22 (b) St. Catharines, 1983

The area that surrounds growing cities changes in land use quickly. Land that was vacant soon becomes built up with houses, factories, and other urban land uses.

Such an area is neither a fully built up city (urban) nor is it a farming area (rural). For this reason this zone is referred to as a **rural-urban fringe.**

Figure 4-23 (a) Calgary, 1968

Figure 4-23 (b) Calgary, 1984

34. Explain in detail the changes that have taken place in St. Catharines and Calgary between 1960 and 1984.
35. (a) In what direction does the city shown in Figure 4.24 appear to be growing? Give reasons for your answer.
    (b) What three specific land uses indicate that this area is not yet fully built up as a city? Explain your answer.
    (c) Give two examples of land uses on the map that are not suitable to be located in a city. Give reasons to explain your choices.
36. (a) Where in this area would you choose to live? Why?
    (b) Identify four locations where you would not want to live. Give reasons for your answer.

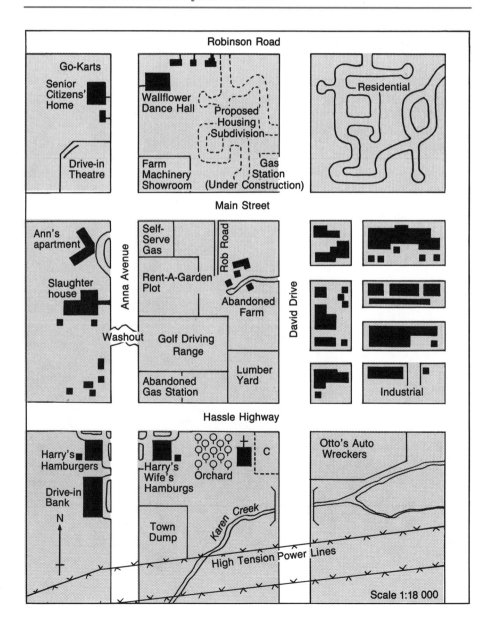

Figure 4-24 Rural-urban fringe

It is apparent from the map showing rural-urban fringe that some land uses are in conflict with each other. To avoid this problem most Canadian cities have set up **zoning laws.** Such laws specify where certain land uses will be permitted. Zoning laws are part of a whole process called **urban planning.** The purpose of planning is to build a city that is best suited to the residents. The planned city should be attractive, well organized, and a generally desirable place to live.

The local or municipal governments in Canada generally have a planning department that tries to ensure that development will be well organized. When a landowner wants to develop a section of land, the development plan is usually submitted to the planning department and municipal government for a decision to approve or disapprove the proposal.

The local governments in Canada also provide important services for a town or city such as water supply, sewers, garbage collection, snow removal, roads, and schools. **Property taxes,** which are collected from every land owner, provide much of the money for these services.

---

37. (a) **What is the name of the local government that provides services for your home?**
    (b) **Which of the services mentioned above are or are not provided for your home?**

MEGALOPOLIS

As cities expand they may actually touch each other. Dartmouth and Halifax in Nova Scotia, or Kitchener and Waterloo in Ontario are excellent examples of cities that have grown to the point where they meet. If several cities do this the new urban area can be considered a megalopolis, which is literally an extremely large city.

In Canada there is an area in which a megalopolis has begun to form. Along the shore of western Lake Ontario there is continuous urban development.

---

38. (a) **Turn to a map of southern Ontario in your atlas. Locate the area referred to above and list, in order, the cities that have either grown together or may soon join this urban development.**
    (b) **Using the appropriate scale, measure the driving distance in kilometres from one end of this area to the other.**
39. (a) **Refer to the agriculture map in your atlas. What specific land use is being taken over by the growth of these cities?**
    (b) **In which ways might this urban growth be considered a problem? What might result a few years from now?**

LAND USES IN A CITY

Within each large Canadian city there is a wide range of land uses. For the purposes of our study we will concentrate on the following three:

- Residential—where people live
- Industrial—factories and manufacturing areas
- Commercial—where people shop and conduct their business

These three categories are simplified, but give a general picture of the structure of the city. One important concept to remember is that the oldest sections of a city tend to be in the central area, or downtown. Often the downtown area is referred to as the **Central Business District (CBD)** and usually has the tallest office buildings of the city.

Cities frequently grow outward in rings, so that the newest areas tend to be on the outskirts. Notice how Metropolitan Victoria has grown by looking at Figure 4.27. The suburbs of Canadian cities often have the newest plazas, widest roads, and most modern facilities. Recently, however, there has been a trend toward **redevelopment** in downtown areas. This process involves the tearing down of old buildings and their replacement with new structures.

Figure 4-25 The central business district of Montreal

Figure 4-26 Redevelopment

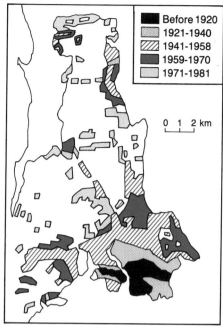

■	Before 1920
	1921-1940
▨	1941-1958
▨	1959-1970
	1971-1981

0 1 2 km

Figure 4-27 Growth of the built-up area of metropolitan Victoria (B.C.), 1920-1981

---

40. (a) Why would newer areas of a city tend to have wider roads and more plazas?
    (b) Why would the tallest buildings of a city be downtown?
41. Place a piece of tracing paper over the photograph of Figure 4.25. Trace the outline of the river and the area of tallest buildings in Montreal. Shade in the CBD and label the St. Lawrence River.

Residential

People in cities live in the type of housing they either prefer or can afford. Their age, size of family, and way of life also help to determine the homes people live in.

---

42. (a) In Figure 4.28, which home is oldest? How can you tell? Which section of the city would it be found in?
    (b) Which types of homes could you expect to find in the suburbs of a city? Give reasons for your answer.
43. What type of housing would each of the following people probably live in? Explain your answers.
    (a) a retired man, age 81
    (b) a married couple with four young children
    (c) a single university student

Figure 4-28 Different types of residential buildings

Although homes of various types appear to be scattered at random across the city, there is usually some pattern to be observed. The more expensive homes occupy the more desirable locations, such as near a river valley, a lake, or a forest. The least expensive homes are often located in less desirable areas, such as near industry.

One important trend that is occurring in larger cities is the drift towards apartment living. Apartments provide homes for many people on a small piece of land. In this way they are ideally suited for increasingly crowded cities where land is very expensive.

Institutional land uses such as churches, schools, and libraries are located in or near to residential areas.

44. In your opinion, what are the advantages and disadvantages of living in an apartment, as opposed to a house with its own lot and garden?

Industrial

Industry can be described as the economic backbone of a city. Without industry a city could not earn the money necessary to exist. Not only does industry provide jobs for the residents of the city, but it also pays relatively high property taxes to the local government for its services. Industrial sites often are located near transportation routes such as highways, bodies of water, or railways. Modern factories also demand a large area of land, so they are frequently located on the outskirts of cities. Planned developments for factories are increasingly popular and are called **industrial parks.**

Figure 4-29 An industrial park and a modern factory

Commercial

Traditionally, the CBD has been the centre of shopping in Canadian cities. The CBD is usually reached fairly easily from various sections of the city.

With the growth of cities and the increasing use of cars, suburban plazas have come into prominence. They are usually located along main roads or highways and are designed for the convenience of the shoppers. With huge parking lots and many stores, these plazas are ''one-stop shopping centres'' for many Canadians today.

Figure 4-30 Inside a modern shopping plaza   **Figure 4-31 A commercial strip**

The popularity of plazas has hurt the stores in the CBD, as well as the smaller ones located throughout the city. Stores found in **commercial strips,** or rows along major streets, for example, have lost much business to indoor plazas.

45. What specific advantages does a large suburban plaza offer shoppers over a commercial strip?
46. When shopping is done for your household, which of the three store locations is used most frequently? Why?

URBAN PROBLEMS

Along with all the benefits and facilities of our cities, there are problems that inevitably develop.

47. (a) Which of the problems shown in Figure 4.32 have you noticed in your city or a city near where you live?
    (b) Under what specific conditions have you noticed the problems shown in photographs (a) or (c)?
    (c) In what section of the city would you likely find conditions such as are shown in Photograph (d)?
48. Secure a recent copy of a city or town newspaper. Examine the various news items and fill in a table of your own based on the one shown on the next page. Include only articles reported from your local area.

(a) Heavy air pollution

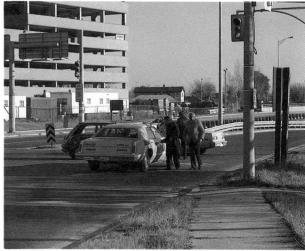

(b) Urban crimes

(c) Traffic jams

(d) Slums

Figure 4-32 Some urban problems

TYPE OF PROBLEM	CRIME	POLLUTION	TRAFFIC PROBLEMS	OTHER PROBLEMS
Specific description of problems (e.g. theft or accident)				
Possible solutions (If no ideas are contained in the newspaper article, make your own suggestion)		SAMPLE	ONLY	

Case Study: Fredericton, New Brunswick

Population: 43 723 (1985). Officially became an urban centre in 1848.

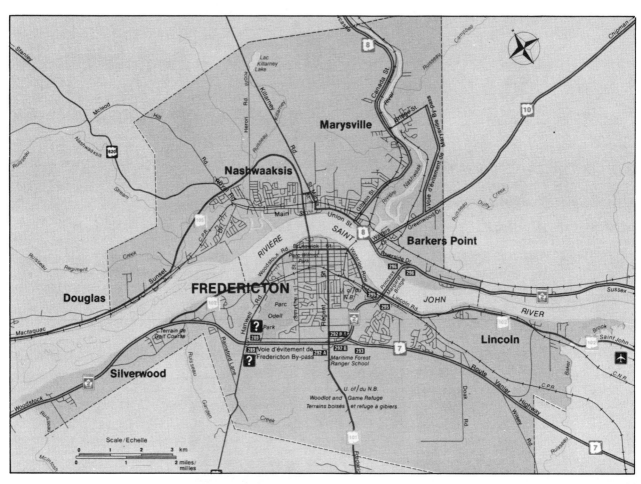

Figure 4-33 (a) A map of Fredericton

Figure 4-33 (b) Downtown Fredericton

Use your atlas to help answer the following questions.

49. (a) Describe the site and situation of Fredericton.
    (b) What appear to be the two most important employers in Fredericton?
50. What specific portion of the map of Fredericton is shown in the photograph?
51. The major streets of Fredericton follow a grid pattern. What advantages does this grid pattern have over the pattern shown in the eastern section of the city?

Case Study: Kamloops, British Columbia

Surveys and Mapping Branch ©
Her Majesty the Queen in Right of Canada with permission of Energy, Mines and Resources Canada.

Figure 4-34 Air photograph of Kamloops

52. (a) Using your atlas, discover and record the approximate population of Kamloops.
    (b) Why do you think Kamloops was founded on this site?
53. (a) What kinds of transportation links does Kamloops have with the rest of British Columbia? To what cities do they lead?
    (b) What port city would include Kamloops in its hinterland?
54. What are the major industries of Kamloops?

Case Study: Toronto, Ontario

In 1792, Lord Simcoe chose the site of present-day Toronto for the establishment of a new town, originally called York. As the governor of Upper Canada (now Ontario), Lord Simcoe made a decision that was to be significant for many years. Although there were other locations where a town could have been founded in southern Ontario, Lord Simcoe's influence was decisive.

Since 1792 Toronto has grown quickly and today Toronto and its surrounding region is the largest metropolitan area in Canada.

There are a number of characteristics about Toronto's situation that have helped it to grow.

---

55. (a) In what ways and for what reasons do transportation routes influence the outward growth of a city?
    (b) What transportation routes may have contributed to the growth of Toronto? What cities do they link with Toronto?
    (c) On a map of eastern North America mark the city of Toronto and then join it with straight lines to all major Canadian and American cities within 1 200 km. Give your map an appropriate title. What significance would this situation have for manufacturers in Toronto?
56. (a) Examine the photograph in Figure 4.36. What makes Toronto a suitable location for a break-of-bulk point?
    (b) What other land uses can you see around Toronto harbour? Why would they be located there?
    (c) Using a map of eastern Canada from your atlas, suggest which Canadian cities would ship goods through Toronto. Describe the route ships would take from Toronto to reach countries other than the United States.

Figure 4-35 The Toronto skyline

Figure 4-36 Toronto Harbour

Thematic maps
Seigneury
Seigneur
Rang
Roture
Surveying
Base line
Concessions
Concession line
Side roads
Section
Farmsteads
Hamlet

Four corners
Village
Town
Economic base
Ghost town
City
Break-of-bulk
Hinterland
Site
Situation
Rural-urban fringe
Zoning laws
Urban planning

Property taxes
Megalopolis
Central Business
   District (CBD)
Redevelopment
Residential
Institutional
Industrial
Industrial parks
Commercial
Commercial strips

Research Questions

1. (a) From your teacher obtain a map of your local area including your school. On that map name each road and mark on and label each land use. Choose a partner to help map your area.

   Use these land-use categories and devise a key, using different colours to represent each of the categories.
   (1) Residential
   • Apartments
   • Single-family detached and semi-detached houses
   • Townhouses (row houses)
   (2) Commercial
   • Shopping, banks, etc.
   • Offices
   (3) Industry
   (4) Parks
   (5) Institutions (label them specifically)
   (6) Farmland
   (7) Streams, creeks, swamps, lakes
   (b) What is the dominant land use of your local area?
   (c) What changes could or should take place in the area near your school? Explain your answer.
2. If you live in a town or city, find out the following information:
   (a) The population of your urban area.
   (b) The number of people on your town or city council.
   (c) Whether your local government has a planning department. If so, determine its function.
   (d) The steps a landowner must take in order to develop a tract of land.

3. (a) When was your local government established?
   (b) Describe the historical development of your local area, referring to changes in land uses and/or population. Include information that you can obtain from your parents, neighbours, long-time residents, or libraries in your area.
4. (a) Describe the site and situation of your village, town, or city, or one close to you. Use maps available from your teacher to help in your answer.
   (b) Why did that urban centre grow up where it is?

5 FARMING IN CANADA

Canadian farms are among the most modern in the world and provide a great variety of high quality food for our dinner tables. At times we take our food for granted and forget the many hours of work that were invested in the production of that food.

The following quiz will help you to begin thinking about Canadian farms. Record the answers in your notebook.

QUIZ on farming in Canada

1. What is the value of an average farm in Canada?
   (a) $ 50 000
   (b) $125 000
   (c) $200 000
   (d) $500 000

2. On the average, where are the largest farms in Canada found?
   (a) the Maritimes
   (b) British Columbia
   (c) the Prairies
   (d) Ontario and Quebec

3. For a broiler chicken to gain 1 kg, how much must it consume?
   (a) 0.5 kg of feed
   (b) 2 kg of feed
   (c) 5 kg of feed
   (d) 10 kg of feed

4. Of each dollar that you spend on food, how much does the farmer receive?
   (a) $0.55
   (b) $0.28
   (c) $0.10
   (d) $0.76

5. What percentage of Canada's total area is farmland?
   (a) 75%
   (b) 48%
   (c) 15%
   (d) 7%

6. For a cow to gain 1 kg, how much must it eat?
   (a) 4 kg of grain
   (b) 7 kg of grain
   (c) 10 kg of grain
   (d) 0.25 kg of grain

7. Which province has the largest area of farmland in Canada?
   (a) Quebec
   (b) Ontario
   (c) Alberta
   (d) Saskatchewan

8. In which province has the most farmland been abandoned in the last 25 years?
   (a) New Brunswick
   (b) Manitoba
   (c) Newfoundland
   (d) British Columbia

9. How many people does each Canadian farmer, on average, produce enough food to support?
   (a) 5 people
   (b) 55 people
   (c) 30 people
   (d) 10 people

10. As a result of serious drought (lack of rainfall) in the Prairies in the 1930s, how many Prairie farmers abandoned their land?
    (a) 1 000 000
    (b) 2 000 000
    (c)   150 000
    (d)   250 000

The answers to the quiz can be found on page 179.

Clearing the Forest

The early European settlers who arrived in Canada were shocked to discover the extent of the forests of our country. These settlers arrived first in the Maritime Provinces and Quebec, and then moved inland to present-day Ontario. You can imagine their surprise as they looked out on forests that extended almost unbroken through all of eastern Canada.

It is likely that the largest forest these settlers had ever seen before was a small woodlot near their home village. Never had they seen such a seemingly endless stretch of trees.

The shock of first seeing the Canadian forest was nothing compared to the job of clearing it for farming! Once a pioneer bought or was given a forested farm, there was little time to rest. That farm had to be cleared to grow crops, or starvation would follow.

It is not surprising that the first European settlers in Canada considered the forest to be their enemy. They believed that the forests gave Canada a cold climate and produced many of the diseases that the pioneers experienced. It became a top priority for these settlers to clear the forests, regardless of the results.

Unfortunately, there was some land that should never have been completely cleared of trees.

Figure 5-1 Clearing the forest

Figure 5-2 Land not suited for agriculture

Figure 5-3 The results of clearing unsuitable land

1. (a) Explain why the land shown in Figure 5.2 should not be cleared for agriculture.
   (b) Explain why the land shown in the photograph in Figure 5.3 should not have been cleared of forest.
2. Why is it difficult to re-establish a forest in an eroded area?

Changes in Canadian Farming

As you noted in the last chapter, the percentage of Canada's population working on farms is declining rapidly. Fewer people are now needed to produce food than in the past. At the same time as the farm population has been declining, the food production from Canadian farms has been going up. This means that the **productivity** of Canadian farms has been improving, since fewer farmers produce more food.

YEAR	WORKERS	TONNES OF FERTILIZERS	MACHINES
1901	90	—	—
1911	65	—	—
1921	50	—	3
1931	48	13	5
1941	48	15	8
1951	33	30	20
1961	25	43	28
1971	18	78	28
1981	15	98	30

NOTE: Figures refer to each 1 000 ha of cropland.

Figure 5-4 Changes in technology on Canadian farms

3. Set up a graph in your notebook using the axes (lines) as shown below.

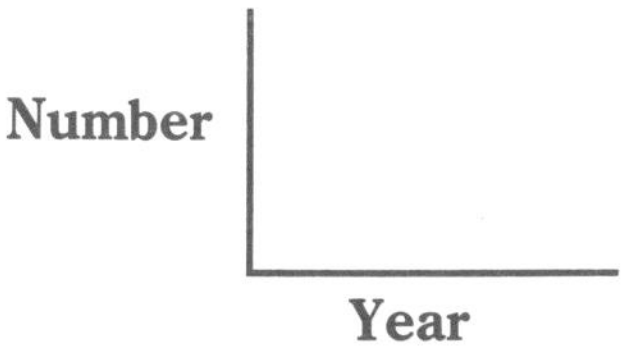

Number |
       |
       |_____
          Year

   Plot a line graph of the number of workers on Canadian farms by year. Label your line. Repeat this process for the other two sets of information.
4. (a) Describe, using your own words, what is shown by the graph you have just completed. What importance do these changes have for the way in which the average Canadian farmer works?
   (b) What industries besides farming have benefited from these changes since 1901?

Figure 5-5 Changes in farming technology

2000	By 2000 it is expected that the first generation of farm robotics (computer operated machines) will be in use.
1990	Embryo transplants with cloned cells produce better plants and animals (genetic engineering). Biological pest controls used in combination with new farm chemicals. Microcomputers increase farm efficiency.
1980	Income stabilization programs reduce financial risks in farming. National marketing plans for milk, eggs, turkeys, and chickens. Artificial insemination.
1970	Computers to help farmers. Machines to harvest potatoes.
1960	Raising of chickens in controlled temperature environment. Lightweight tractors.
1950	New, effective herbicide (weed killer) 2-4-D introduced. New pesticides introduced.
1940	New combine machines cut, thresh, clean grain. Frozen foods.
1930	Homogenized milk.
1920	Boom in manufacturing of new farm machines. Motor trucks appear on farms.
1910	New Marquis wheat introduced, ready to harvest in one hundred days. Powdered milk produced.
1900	Refrigerated railway cars.
1890	Canadian Pacific Railway carries produce across Canada. New Massey lightweight binder for grains.
1880	New grain harvester made by the Massey Co.
1870	First steam-powered plows.
1860	Hand tools still used.
1850	

It should be noted from the ladder of progress that in 1850 hand tools were still in use on Canadian farms. These tools had changed little in hundreds of years. Since 1850 the changes in farming methods have been greater than those in the thousand-year period before 1850.

---

5. How would your life be different today if the changes shown on the ladder had not taken place? Mention specific examples in your answer.
6. One problem farmers have always had is the spoiling of food. Which changes have helped to eliminate the spoiling of food?

The Importance of Agriculture

Every day Canadians spend over $100 000 000 on food for home consumption. Even though this figure does not include the money Canadians spend at restaurants, it does show the great importance of farm produce in our daily lives. Where would we be without farms?

Agricultural produce is also a major export for Canada. Grains such as wheat, barley, and oats help to generate a trade surplus in agricultural goods of almost $10 billion per year for Canada. Although Canada imports a great deal of food, productive Canadian farms produce and export more food than is imported. In fact, 11% of all Canadian exports are agricultural in origin at the present time.

In addition, Canadian farms are very important because of the number of people who live and work on them. There are over 300 000 farms in Canada. If an average farm supports four people, you can see that well over 1 000 000 people depend on farming for their livelihood.

---

7. (a) Check the population of Canada given in Chapter 1 and then calculate how much the average Canadian spends on food per day and per year.
   (b) What goods can you buy that are equal in value to the amount of money you might spend per year on food (as in (a))?
   (c) What does this show you about the price of food as an essential for life in Canada?

FROM THE FARM TO YOUR TABLE

Many industries depend upon agriculture for their existence. Hundreds of thousands of Canadians work in the railways, food processing plants, supermarkets, and storage facilities so essential to the movement of food from the farm to the consumer.

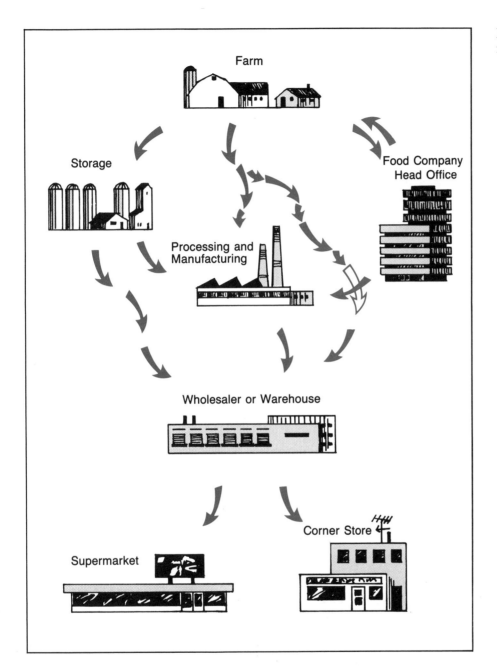

Figure 5-6 The food marketing system

8. When you buy food at a supermarket or corner store, you probably do not think of all the stages it passed through before reaching you. From the diagram (Figure 5.6), list the various steps that the food passes through in which cost is added. Two examples would be trucking and storage costs.

9. Today many Canadians are concerned about the various artificial chemicals and preservatives added to food. Examine the whole process shown in Figure 5.6 and explain why you think such preservatives are considered necessary.

Distribution of Different Types of Agriculture in Canada

The type of agricultural activity that develops in any area depends mainly on the climate and soils. Where very special climatic or soil conditions exist, the land may be used for crops like vegetables or fruit. Vegetable and fruit farms are usually quite small, yet they require much labour to operate. Such farming is called **intensive farming**. Wheat farms and ranches, on the other hand, are usually very large and involve less labour. Wheat farming and ranching are types of **extensive farming**.

Figure 5-7 Distribution of different agricultural types in Canada

10. What relationship can you discover between the location of agriculture in Canada (Figure 5.7) and Canada's population distribution (Figure 4.1 on page 113)? Explain the reasons for this relationship.

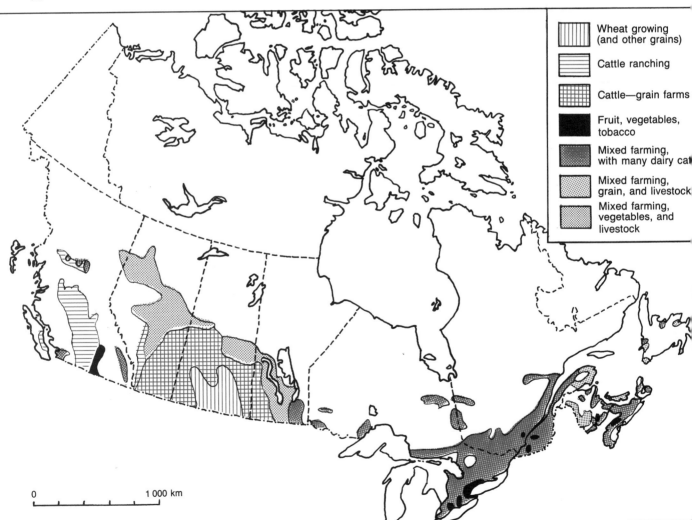

Wheat growing (and other grains)

Cattle ranching

Cattle—grain farms

Fruit, vegetables, tobacco

Mixed farming, with many dairy cat

Mixed farming, grain, and livestock

Mixed farming, vegetables, and livestock

0   1 000 km

11. (a) What type of farming occupies the least amount of land in Canada? In what parts of which provinces is it found?
    (b) Which farming type extends farthest north in Canada? In which province does this occur?
12. Using Figures 1.3 (page 3) and 5.7, list the regions of Canada that have almost no farming. Suggest reasons to explain this.

WHEAT FARMING

Canada is one of the major wheat exporting countries of the world. In fact, the Canadian Prairies are considered the bread basket of Canada, since most of our wheat is grown there. The Prairies are well suited for the cultivation of wheat. With rich grassland soils, cool wet spring weather, and hot dry summers, the Prairies provide wheat with generally excellent growing conditions. Long periods of **drought,** when little or no rain falls, cause the plants to die and the developing grain to shrivel.

Prairie farmers must face a series of other problems besides that of dry weather. In most cases wheat requires a minimum of about 100 days without a heavy frost to mature and ripen. Early frost or snowfalls can seriously damage a wheat crop. Wheat diseases, such as rust, can also hurt grain production. New resistant **strains** (varieties) of wheat have helped to reduce the danger of damage from disease.

Weeds and pests such as grasshoppers and cutworms can seriously harm Prairie crops. With modern **pesticides** to kill animal pests and **herbicides** designed to control weeds, the farmer has succeeded in countering these problems to some extent.

Figure 5-8 Harvesting wheat in the Prairies

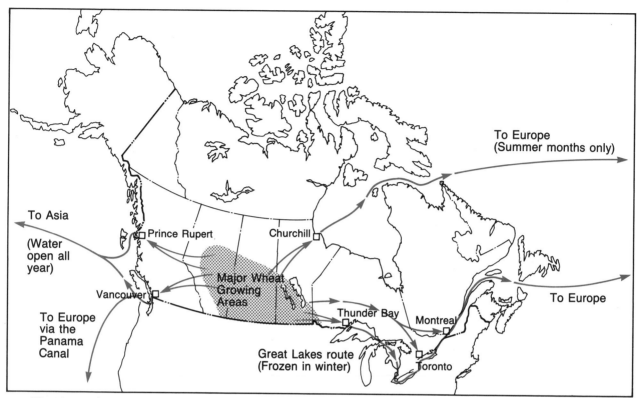

NOTE: Wheat farmers depend upon an efficient transportation network to get their crop to the customers. Trains and ships move the grain from local elevators to destinations all over the world.

Figure 5-9 Wheat transportation routes

13. In what ways is wheat farming important to many other areas of Canada?
14. Imagine that you are a Prairie wheat farmer, and try the following activity.

Wheat Farming: A Gamble

This game should be played in groups of four or five. Each player will need a small marker and each group must have a die. The aim of the game is to reach September 22 first, for a successful wheat harvest. Each player rolls the die and the one with the highest number starts. Each player in turn rolls the die and moves his or her marker forward the appropriate number of spaces. If you land on a 🌾 square, it indicates good wheat growing conditions. Look in the "ideal conditions" column on the left to find out how many days to move forward. If you land on a 🌾 square you have met with a problem. Refer to the "problem conditions" column to find out how many days to move back. The player who first reaches September 22 is the winner, having been the first to successfully harvest his or her crop.

15. Write half a page to describe why wheat farming is a gamble.

156

THE CALENDAR FOR PLAYING WHEAT FARMING: A GAMBLE

PROBLEM CONDITIONS	IDEAL CONDITIONS	CALENDAR							FARMING ACTIVITIES
FLOODING (-2)	WARM, DRY (+2)	START MAY 20	21	22	23	24	25	26	PLANTING
LATE FROST (-2)		27	28	29	30	31	JUNE 1	2	
DROUGHT (-2)	WARM, MODERATE RAINFALL (+3)	3	4	5	6	7	8	9	GERMINATION
		10	11	12	13	14	15	16	
		17	18	19	20	21	22	23	
INSECT DAMAGE (-1)	HOT, SUNNY, FAIRLY DRY (+2)	24	25	26	27	28	29	30	GROWING
		JULY 1	2	3	4	5	6	7	
TOO MUCH RAIN (-1)		8	9	10	11	12	13	14	
		15	16	17	18	19	20	21	
DISEASE (-1)		22	23	24	25	26	27	28	
		29	30	31	AUGUST 1	2	3	4	
TOO MUCH RAIN (-1)		5	6	7	8	9	10	11	
HAIL/WINDSTORMS (-3)	HOT, DRY (+3)	12	13	14	15	16	17	18	RIPENING
		19	20	21	22	23	24	25	
DROUGHT (0)		26	27	28	29	30	31	SEPTEMBER 1	
EARLY SNOWFALL (-4)	WARM, DRY (+4)	2	3	4	5	6	7	8	HARVESTING
		9	10	11	12	13	14	15	
POOR MARKET (-2)		16	17	18	19	20	21	22 FINISH	

157

Figure 5-10 Rounding up the herd

RANCHING

When most Canadians think of ranching, they imagine a rugged-looking cowboy riding a horse across a wide open field. The cowboy movies, which are so much a part of our entertainment industry, have made us very aware of ranching. On television the cowboy shows are set in the "Wild West," where gun battles, lawless towns, and bank robbers were the order of the day.

The Canadian west, however, was not the Wild West of the cowboy movies. In fact, our western provinces were settled in an orderly, lawful way, far different from the settlement of the western United States.

The cattle ranching that began with the settlement of the Canadian west still continues today. Some of the present-day cowboys are Indians. Whatever the cowboys' background, they have a tough life to follow, with long days and hours of riding with the cattle. Even today, with airplanes to track the cattle on the ranches, horses are still widely used. Riders on horses can reach places in rough country where modern machines could not.

Large-scale ranching takes place in Alberta and British Columbia. In Alberta the ranches are on rolling grasslands called the foothills of the Rocky Mountains. Calgary is located in this ranching country and every year it hosts Canada's largest rodeo, the Calgary Stampede. During Stampede time the cowboys engage in contests that range from steer-roping to chuck-wagon races.

Figure 5-11 Scenes from the Calgary Stampede

The second large ranching region is in the Interior Plateau of British Columbia. Here the land is much rougher than near Calgary. In some cases ranches extend up an entire mountainside. On such large ranches the cowboys and the cattle would move up and down the mountain according to the season. This would ensure that they reached fresh pasture throughout the year. This seasonal movement of people and animals is called **transhumance.**

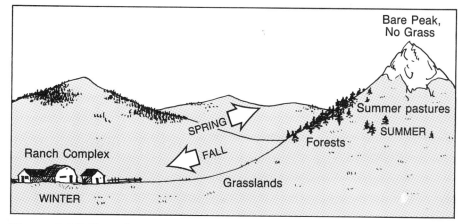

Figure 5-12 Transhumance

Figure 5-13 A feedlot

In Alberta and central British Columbia, the calves are raised until ready to be transported to markets in other parts of Canada. Increasingly, cattle are taken to **feedlots** near cities, where they are fattened with grain before slaughter.

On each ranch all the steers are branded with the ranch's symbol to allow easy identification if a steer ever escaped. Barbed wire is used extensively on western ranches, but breaks in the wire frequently occur. It is a constant job for the ranch owners and cowboys to survey the fences for breaks, then repair them.

16. (a) Describe what actually happens with the process of transhumance in British Columbia. Use Figure 5.12 as a guide.
    (b) In what specific ways is ranching a suitable form of agriculture for the rough country of western Alberta and British Columbia?
17. What particular skills would be important for a modern cowboy to have? Explain the reasons for your answer.

DAIRYING

Milk and its products are an important part of almost every Canadian's diet, as show in Figure 5.14.

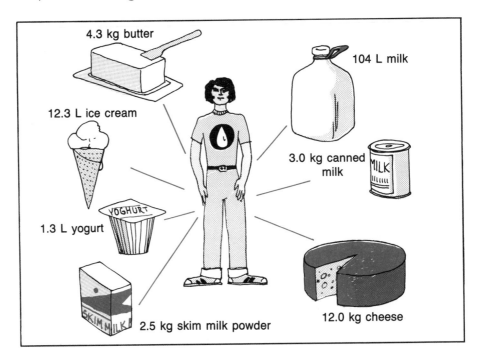

Figure 5-14 Average annual consumption of dairy products by each Canadian

While dairy cattle are kept in almost all parts of the country, certain areas predominate. In 1982 Quebec produced nearly 74 864 kL of milk, well over one-third of the total Canadian production. Much of this is used as *industrial milk,* which includes all products except fluid milk. Ontario dominates in the production of milk for fluid consumption. Dairying is also important in Alberta and British Columbia.

Case Study: Operating a Dairy Farm

Operating a dairy farm is probably one of the most demanding farming activities. Most dairy farmers rise early, at about 04:30 h in the morning. One such farmer is Mr. Guillaume Cooke. With the help of his employee, Pierre, he brings his 31 cows to the barn from the field. In the winter, cows may spend all day in the barn. Pierre prepares the animals for milking. He ties them up, gives them grain to eat and cleans their *udders*, the milk producing glands. While the cows are being prepared Mr. Cooke passes a chlorine solution through the milking pipeline to sterilize it.

Mr. Cooke's equipment allows him to milk three cows at a time. The whole process takes just over one hour. A good milking cow will produce about 50 L of milk each day.

The milk passes through the pipeline to a large storage tank at the end of the barn. This tank, which is kept cool, is emptied into a special milk truck once every two days. The truck then takes the milk to a dairy processing plant.

Figure 5-15 Milking
Dairy cattle in the barn, waiting to be milked.

18. How much milk would be produced in total each day?
19. In what ways does the dairy farmer's life differ from that of a farmer who keeps no animals?

In the summer, after all of the cows have been fed and milked they are returned to graze in the fields. In winter the farmer will usually put the cows into the field for one hour each day to give them some exercise.

Mr. Cooke and Pierre then have breakfast. During the remainder of the day the barn is cleaned, feed is prepared for the evening milking, and any fieldwork or repairs are carried out. Evening milking starts at 17:00 h.

Mr. Cooke's farm has an area of 110 hectares. His land is used in the following way:

- 5 ha are planted with *corn*. The corn is cut and stored in a *silo*, a tall structure in the shape of a cylinder. It is used to feed the young, non-milking animals.
- 55 ha are planted with *alfalfa*, a low leafy plant. It is cut, dried, bailed, and stored in the upper part of the barn. Alfalfa makes good winter feed for the cows. The cows are allowed to graze on the alfalfa fields that have been reharvested.
- 40 ha are planted with a mixture of *oats* and *barley* seed. The grain that results is used year-round to feed the animals.
- The remaining land is used for *improvised pasture* and farm buildings. Improved pasture is grassland that has been planted.

---

20. Design your own dairy farm.
    (a) Draw a rectangle on a piece of paper. The rectangle should be 10 cm from left to right and 22 cm in length. This represents your farm boundary.
        Each cm² represents 1 ha. Also 1 cm represents 100 m.
    (b) Draw a road 1 cm wide along the bottom edge of your farm.
    (c) Design your own dairy farm using information from Mr. Cooke's farm as much as possible. Farm buildings should include the following:
        (i)   a farm house
        (ii) a garage
        (iii) a large milking and storage barn
        (iv) a barn for young animals
        (v)   a cement area attached to the barn
        (vi) a storage shed for tractors and other equipment
    Include a laneway to the house and the barn for the milk trucks and farm vehicles.

Running a farm involves large sums of money. To buy a farm such as Mr. Cooke's, with its buildings, equipment, animals, and milk quota, would cost about $1 000 000. Each year Mr. Cooke receives about $120 000 from milk sales, but his expenses amount to about $75 000, including $25 000 in wages, $6 500 for food additives, and $4 300 in taxes. A dairy farmer not only has to know how to look after animals, but also to keep accounts carefully and be a fairly good mechanic.

---

21. (a) Assume that Mr. Cooke sold his farm and invested the $1 000 000 he received. If the interest rate on his investment was 10% each year ($0.10 earned on each dollar) what would his annual income be?
    (b) How does this income compare to what he is earning at present?
22. In view of your answer to 21 (b), why would a person like Mr. Cooke continue to farm?

ORCHARDS

Although Canada is a country with vast stretches of farmland, there are only four major areas where fruit is grown. To ensure a good crop of fruit, certain special weather and soil conditions must be met. As a result of these specialized growing conditions, most of Canada is unsuitable for fruit farming.

One important condition necessary for fruit growing is a moderate climate with mild winters, hot summers, and few **killing frosts.** Such frosts occur when temperatures drop below 0°C during the fruit growing season and severely damage the buds, blossoms, or fruit. Each of Canada's major orchard regions is located in such a way as to reduce the possibility of killing frosts.

AREA	POSITION RELATIVE TO WATER	OTHER IMPORTANT FEATURES
Okanagan Valley, British Columbia	Near Lake Okanagan for irrigation water.	All of these areas are on slopes which allow cold air to drain away. They also have well-drained, light soils suitable for fruit trees.
Niagara Fruit Belt, Ontario	Near Lake Ontario for mild winter temperatures.	
Southern Quebec	Not near a large body of water.	
Annapolis Valley, Nova Scotia	Near the Minas Basin and the Annapolis River.	

Figure 5-16 Important characteristics of our fruit growing areas

23. Describe what an ideal location for fruit growing in Canada would have to be like. Base your answer on the table (Figure 5.16).

Wherever the fruit farm is located in Canada, the cycle of activities resembles that shown in Figure 5.17.

24. Use Figure 5.17 to help you to answer these questions.
    (a) During which month(s) could the grower take a holiday? Explain why this is true.
    (b) During which month(s) is there most activity on the farm?
    (c) When would the farmer need the greatest amount of hired labour to help? What problems would a farmer have in trying to obtain labour for just that period of time?
25. Write a short note of half a page to describe the three major activities that an orchard farmer would be involved in during the year.
26. What effect would there be on a fruit farm if there was no work done in April at all?

Figure 5-17 Year-round fruit farming activities

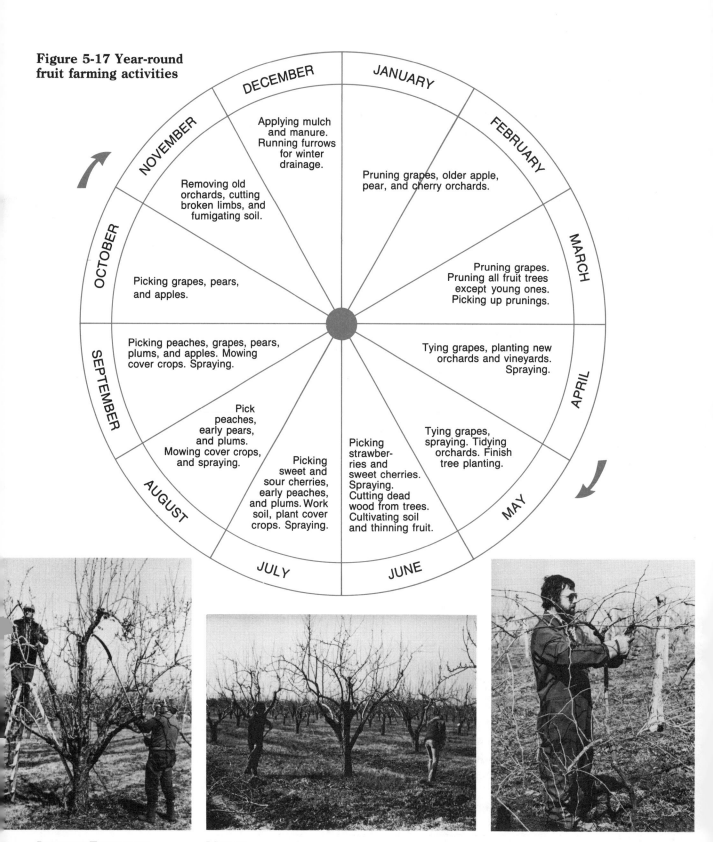

DECEMBER
Applying mulch and manure. Running furrows for winter drainage.

NOVEMBER
Removing old orchards, cutting broken limbs, and fumigating soil.

OCTOBER
Picking grapes, pears, and apples.

SEPTEMBER
Picking peaches, grapes, pears, plums, and apples. Mowing cover crops. Spraying.

Pick peaches, early pears, and plums. Mowing cover crops, and spraying.

AUGUST
Picking sweet and sour cherries, early peaches, and plums. Work soil, plant cover crops. Spraying.

JULY

JUNE

JANUARY
Pruning grapes, older apple, pear, and cherry orchards.

FEBRUARY

MARCH
Pruning grapes. Pruning all fruit trees except young ones. Picking up prunings.

APRIL
Tying grapes, planting new orchards and vineyards. Spraying.

MAY
Tying grapes, spraying. Tidying orchards. Finish tree planting.

Picking strawberries and sweet cherries. Spraying. Cutting dead wood from trees. Cultivating soil and thinning fruit.

JANUARY-FEBRUARY   **MARCH**

164

APRIL

MAY

JUNE

SEPTEMBER

OCTOBER

NOVEMBER

JULY TO NOVEMBER

165

Case Study: The Okanagan Valley

Located in the central section of the Western Mountain Region, the Okanagan Valley has the driest location of all the major fruit growing areas of Canada. With an average annual precipitation of approximately 300 mm, the orchards must be **irrigated** (artificially watered). This irrigation water is generally provided from lakes above the valley. The water flows down the valley in **flumes** (troughs) and is then sprayed in the orchards. The sprinklers are short so that their water is sprayed under the trees.

Figure 5-18 Irrigating fruit trees

Figure 5-19 Cross section of the Okanagan Valley

Figure 5-20 The Okanagan Valley

---

27. (a) Examine Figures 5.19 and 5.20. From what part of the cross section was the photograph (Figure 5.20) taken? Explain the reasons for your choice.
    (b) What indications are there from Figure 5.20 that the area is very dry?

MARKET GARDENING
Case Study: Holland Marsh

Imagine a place where each hectare of land could produce 60 000 kg of carrots, 1 000 crates of celery, 45 000 kg of onions, or 25 000 heads of lettuce. Also consider the value of this productive land if it was close to an area of dense population.

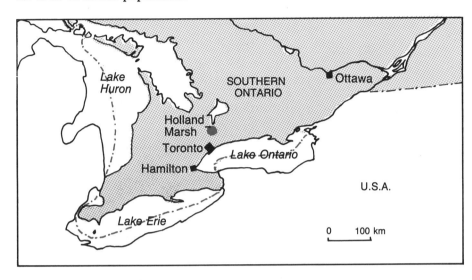

Figure 5-21 Location of Holland Marsh

Holland Marsh, about 50 km north of Toronto, is such an area. It produces large quantities of perfectly formed vegetables each year. It supplies the needs of the many large cities and towns of southern Ontario and other parts of eastern Canada. Some produce is even exported to Norway, Japan, and the West Indies.

About 60 years ago Holland Marsh was just a swampy area; all that was harvested was marsh hay. This hay was used to stuff mattresses and horse collars. For the harvest, wide flat boards were strapped onto the horse's hooves to stop them from sinking into the soggy ground.

Figure 5-22 Horses being used to harvest marsh hay

A scientist realized that the marsh could be made into productive land if it could be drained. Between 1925 and 1930 an area of land 11 km by 3 km was drained.

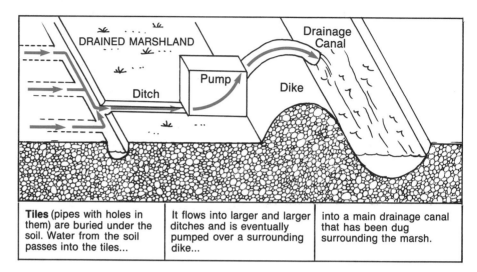

DRAINED MARSHLAND

Drainage Canal

Pump

Ditch

Dike

Tiles (pipes with holes in them) are buried under the soil. Water from the soil passes into the tiles...	It flows into larger and larger ditches and is eventually pumped over a surrounding dike...	into a main drainage canal that has been dug surrounding the marsh.

Figure 5-23 The Holland Marsh system of drainage

The whole system of waterflow can be reversed, when necessary, to irrigate the land. When the correct types of fertilizers were added to the dark "muck" soil, it formed a very productive soil.

Figure 5-24 An air photograph of Holland Marsh

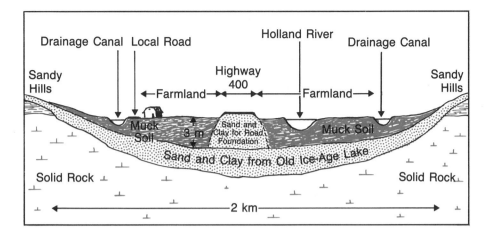

Figure 5-25 A cross section through Holland Marsh

28. What evidence is there from the air photograph that the farms are small and the farming is intensive?
29. Put a sheet of tracing paper over the air photograph and hold it in place with paper clips. Draw and name the following:
   • the area of farmland in the Marsh   • the drainage canal
   • the Holland River   • roads and houses
30. Referring to the cross section,
   (a) explain why the hills surrounding Holland Marsh are not suited to intensive farming
   (b) describe and explain the preparations that were made before the highway was completed.

In 1931 many Dutch settlers from other parts of Ontario were encouraged to come to the marsh. They were familiar with using the drained soils of their native Holland and worked very hard to clear and cultivate the land. Soon the settlement of Ansnorveld was established, and immigrants from many different nations joined the original Dutch settlers after the Second World War.

Figure 5-26 Ansnorveld

Since the marsh was first drained the types and areas of different crops have changed considerably.

CROP	1954	1974	1978	1982	1985
Carrots	19	33	31	40	40
Onions	21	31	34	33	33
Lettuce	29	12	12	10	9
Potatoes	19	11	9	4	5
Celery	7	4	4	4	4
Parsnips	0	1	2	1	1
Cabbage	1	2	1	1	1
Cauliflower	2	2	2	2	2
Beets	1	1	1	1	1
Other crops	1	3	4	4	4
Total hectares cultivated	2 659	3 211	3 559	3 715	3 817

Figure 5-27 Changing percentages of production of various crops in Holland Marsh

Use the data from Figure 5.27 for these questions:
31. (a) Which crop covered the largest area of land in 1954?
    (b) What percentage of the land was occupied by this crop in 1985?
32. (a) Which two crops are now of greatest importance in the Holland Marsh?
    (b) These two crops provide a more dependable, year-round income for the farmers than many of the other crops listed. Why do you think this is the case?
33. Using the figures from 1954 to 1985, draw a divided bar graph to illustrate the changing percentages of various crops grown in the Holland Marsh. Figure 3.29 (page 106) is an example of a divided bar graph.

Figure 5-28 Harvesting celery

Despite this great success story, Holland Marsh has one very significant problem. The valuable soil layer is becoming thinner. The reasons for this are listed as follows:

(1) The soil is compacting as a result of the natural water being drained out.
(2) The top layers of the soil are thinning and breaking down. This results in a slow decline in the quality of the soil.
(3) Wind and water blow or wash away the surface layers of the soil.

New methods have been introduced to increase the productive life of Holland Marsh to 150 years. These include reducing the amount of cultivation and carefully controlling the amount of water in the soil.

---

34. (a) List five factors which might control the yield of crops in Holland Marsh. These factors must be within the control of a farmer and be affordable. The use of fertiliser could be one for you to choose.
    (b) Imagine that you are an agricultural scientist, working at Holland Marsh. Choose any two of the crops listed in Figure 5.27. Devise an experiment to find out the effects of the five factors that you listed in 34 (a) on the two crops which you have chosen. Your answer should include a map to show how you would lay out your experimental growing areas.

TOBACCO

Tobacco is an extremely valuable non-food crop, grown mainly in southern Ontario, but also in Quebec and the Maritimes. During the past few years many people have become aware of the health hazards associated with smoking. Smoking is also becoming very expensive. As a result there has been a significant reduction in the amount of tobacco being smoked in Canada. This in turn has led to a decrease in the number of cigarettes that are manufactured.

There is a surplus of tobacco on the market. The farmers are growing more tobacco than the manufacturers can use. As a result, the price that the farmers are paid for their product is decreasing as they compete against each other for sales. Some growers claim that they are losing 50 cents on every kilogram that they sell. Farmers are keeping their own expenses to a minimum. This results in reduced sales of general and specialized equipment, labour costs, as well as a reduction in normal household expenses. This has had a ripple effect throughout the community expecially in southern Ontario, where 90 percent of the crop is grown.

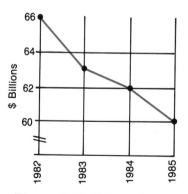

Figure 5-29 Value of sales of Canadian-made cigarettes in Canada

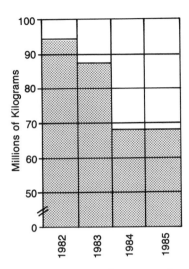

Figure 5-30 Amount of tobacco grown in Ontario

35. List five kinds of activities or people who would be affected by the current crisis affecting the tobacco industry in Canada. Briefly explain how each has been affected.

Tobacco is grown in areas of high summer temperatures and well drained sandy soils. Experiments have been conducted to determine if there are other crops which could grow profitably under the same conditions. These experiments have included peanuts and hardy varieties of peaches, but so far, no crop has been found to grow as well as tobacco.

36. The problems faced by the tobacco industry are an example of market forces at work. Draw a flow chart using arrows to link the following events in the correct order. Use the title "The Effects of Market Forces on the Tobacco Industry."

    UNEMPLOYMENT INCREASES
    LESS TOBACCO IS GROWN
    PUBLIC CONCERN ABOUT HEALTH
    LESS EQUIPMENT IS MANUFACTURED
    FEWER CIGARETTES ARE MANUFACTURED
    ECONOMY OF THE WHOLE AREA IS AFFECTED
    HEALTH CARE COSTS DECREASE
    FARMERS BUY LESS EQUIPMENT

37. In a few sentences explain how Canadians will suffer from and also benefit from the current crisis in the tobacco industry.
38. Imagine yourself as a government official. What would you suggest should be done to improve the current situation in the tobacco growing areas of Canada? Make at least three suggestions.

MIXED FARMING

The types of farming examined so far have all specialized in one crop or a group of similar crops. There are, however, a large number of farms that raise a variety of crops and livestock. Such a farm is called a mixed farm.

39. Refer to Figure 5.7 on page 154 and describe the location by province of the mixed farming areas of Canada.
40. (a) In which province(s) does mixed farming appear to be the dominant type of farming?
    (b) Examine the soil map in your atlas and describe the problem of mixed farming in this (these) province(s).

172

Mixed farming involves raising not only animals but also field crops. These are often fed to the livestock so that a farmer does not need to buy fodder from another source. Corn, for example, may be grown by a farmer to feed his own cattle.

There are a number of advantages to mixed farming. As prices for different farm products vary up and down, the mixed farmer has some stability in income. As one product decreases in price, there is likely another one that increases in price to counterbalance.

Mixed farms can also produce crops that help supply the farmer's own needs as well as those of neighbours. In this way such a farm can be more self-sufficient in food.

As in other types of agriculture there can be poverty on mixed farms. In some cases, the farmer's expenses and income can be almost equal. This leaves little money to buy food, clothing, machinery, or other necessities for the farm. This problem of rural poverty is more complex than it would appear at first. There are several causes; each is difficult to try and solve.

One central factor that contributes to rural poverty is the quality of farmland. Canada has a great deal of marginal farmland where the soil can produce only fair-to-poor crops.

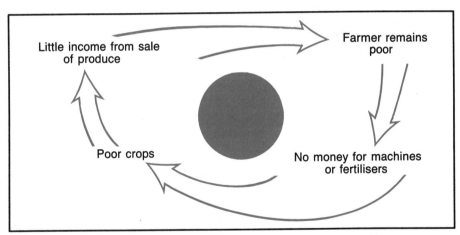

Figure 5-31 The cycle of poverty for some Canadian farmers

NOTE: A poor farmer may have too little training to allow him to move to a non-farming job.

41. Explain in your own words how a Canadian farmer can become trapped in poverty.
42. The map (Figure 5.32) shows a portion of Canada where the land is classified according to its quality for farming. The land has been divided into 7 categories, as has been done in all parts of Canada.
    Category 1 represents the best farmland with good soil, no steep slopes, and good drainage.
    Category 7 represents poor land unsuitable for farming, perhaps because of rocky soils or steep slopes.

Figure 5-32 The area available to you for placing your farm

1 km

Where to Place Your Farm

Instructions:
(a) Locate your farm on the map. Do not mark the book. Copy the details of your farm into your notebook using tracing paper.
   (i) Your farm should be 1 km².
   (ii) Your farm should have as much good land as possible.
(b) Explain the advantages that your farm has.
(c) What problems might you meet in trying to farm the land you have chosen?
43. Describe two difficulties you encountered in trying to choose the best farm site.

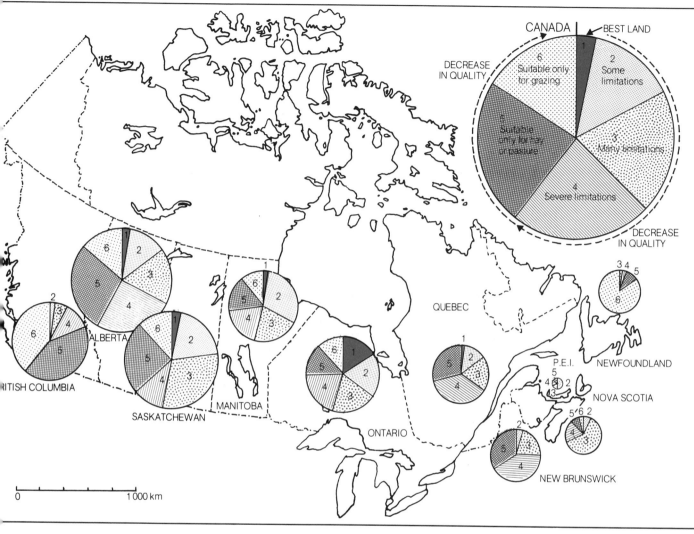

CANADA

BEST LAND

6
Suitable only for grazing

2
Some limitations

DECREASE IN QUALITY

5
Suitable only for hay or pasture

3
Many limitations

4
Severe limitations

DECREASE IN QUALITY

Figure 5-33 Distribution of Canada's food land ranked by classes of soil

44. (a) Rank the provinces in order by area of farmland, starting with the province that has the largest area available for farms.
   (b) Group the provinces into
      (i) those with the most agricultural land
      (ii) those with the least agricultural land
      (iii) those with intermediate amounts of agricultural land.
45. (a) Which province has the greatest proportion of poor land (Classes 5 or 6)?
   (b) Which province has the largest area of the best land (Class 1)?
   (c) In what ways would land quality influence the economic activities of these provinces?
46. Write one-half to one page describing the distribution of different qualities of agricultural land in Canada.

THE PIONEER FRINGE

As you have already discovered, Canada's early European setttlers became pioneers in clearing our forests and farming the land. They faced the obstacles of harsh weather, illness, isolation, and physical discomfort. Through determination and extraordinary effort, these pioneers settled in most of southern Canada.

Today in Canada there are still pioneers who are opening up new land for farming and settlement. In most cases, this land is on the fringe of presently settled land. In contrast to Canada's first pioneers, these people have access to modern machinery. Bulldozers, tractors, and chain saws, for example, are used in clearing trees. Despite this modern equipment, modern pioneers may still sacrifice conveniences such as electricity, television, and nearness to facilities.

Figure 5-34 Fort Nelson, a small community in northern British Columbia

In Newfoundland modern pioneers have been clearing forests for farmland. The Peace River region of northern Alberta and British Columbia is another example of a pioneer fringe area. In both cases the climate is marginal for farming. During warm years the farmers are able to harvest a substantial crop. If the weather is cold, a crop may be covered by an early snowfall or be killed by frost before the harvest.

For many of these people, life on the pioneer fringe may be too discouraging or too costly. In these cases the pioneers abandon their homes and farms to return to southern Canada. Those who do remain to work in the pioneer fringe risk their future with every crop they plant. The beautiful and rugged landscape, however, still challenges these pioneers to stay and work.

Figure 5-35 The Peace River country

47. Why do you think someone would move to a pioneer fringe area of Canada to farm? Suggest at least four reasons.
48. (a) Locate the Peace River region on a map in your atlas. Describe its location relative to Canada.
    (b) Using the thematic maps in your atlas, describe the natural vegetation, climate, and type of farming found in the Peace River country.
    (c) Would you ever move into such an area as is inhabited by modern-day pioneers? Explain your answer.
49. On a blank sheet of paper, design a newspaper advertisement to attract settlers to the Peace River country. Assume that the land is to be auctioned by the government.

Rural Poverty

Rural poverty is related to a large number of causes, which include the following:
(1) Poor soil
(2) Unsuitable climate (example: too cold, too dry, too wet)
(3) Old-fashioned farming techniques
(4) Farms that are too small for a family to grow enough produce to make a living
(5) A location too far from markets
(6) Low prices for farm produce

Figure 5-36 Rural
poverty

GOVERNMENT PROGRAMS

The federal (national) and provincial governments are very concerned
about the problems of poverty on our farms. As a result, they have
developed numerous programs to try and help those caught in rural
poverty. Provincial governments, for example, have experimental farms
where new crops and farming methods are worked on.

The federal government has set up a number of programs as well.
ARDA (Agricultural Rehabilitation and Development Administration) is a
well-known program sponsored by the federal government. Under ARDA
projects, work has been done in such areas as water and soil
conservation, improvement of farmland, and assistance for rural
economies.

The Farm Improvements Loans Act helps farmers to get back loans
of up to $100 000. These loans must be repaid within 10 years, or 15
years, if the money was borrowed to buy land.

A third government program is called the PFRA (Prairie Farm
Rehabilitation Administration) and was begun during the serious droughts
of the 1930s. Its prime goal is to develop water-storage systems on the
Prairies, as well as set up irrigation projects.

Since most Canadians live in large cities or towns, they are often
unaware of the great importance of agriculture. Not only does it produce
our food; it employs—directly or indirectly—a large proportion of our
population. Therefore, the problems that farmers must face relate
directly to our lives, and it is important that we should have a basic
understanding of what these problems are.

The research questions that follow will help you to understand
agriculture more fully.

Productivity
Intensive farming
Extensive farming
Drought
Strains

Pesticides
Herbicides
Transhumance
Feedlots
Killing frosts

Irrigated
Flumes
Tiles
Marginal farmland

Answers to the Agricultural Quiz on pages 146 and 147.

1 (d); 2 (c); 3 (b); 4 (b); 5 (d); 6 (b); 7 (d); 8 (a); 9 (b); 10 (d).

1. USING YOUR LIBRARY TO GET INFORMATION

In your local or school library you will find a *card catalogue*. The contents of the card catalogue are arranged in alphabetical order and include book titles, authors, and topics, together with a *call number*. This number will help you to locate the book on the shelves.

For example, if you wish to find out about the growing of potatoes you should look under *Potatoes, Vegetables, Agriculture,* and also *Canada—geography.*

Other important sources of information include the encyclopedias and atlases kept in the *reference section*. Usually, these books may not be removed from the library. There are atlases of Canada and atlases of each province, showing distributions of many items, such as vegetable-growing or even potato-growing areas. Atlases also contain some statistical information in figure or graph form.

To get the most up-to-date statistical information about Canada, look in the latest *Canada Yearbook* or *Canada Handbook,* which should also be in the reference section.

A collection of pamphlets and newspapers may be found under subject headings in the library *vertical file*. These are arranged in alphabetical order and may provide valuable information. These topics may also be listed in the card catalogue.

Many libraries also include cassettes, filmstrips, and films. These materials may also be useful in providing the information you are seeking.

2. WHEN YOU WORK IN GROUPS

(a) Divide up your research so that each person is responsible for obtaining certain parts of the information. Get photographs and diagrams wherever you can.

(b) When you have completed your research, meet together to make sure that your information is complete and fits together well.

3. WHEN YOU HAVE TO GIVE A GROUP PRESENTATION TO THE CLASS

After you have completed your research:

(a) Divide up your presentation. It would be best if each group member presents the part of the topic which he or she has researched.

(b) Make your presentation clear and short. Use the blackboard when it will help to explain your information. Use as many photographs as possible in your presentation.

(c) Remember that you know much more about your topic than other people in your class. You should therefore explain everything very carefully and clearly.

Research Questions

1. The improvement in wheat yields in the Prairies is mainly due to the development and use of better strains of wheat.
Find out
(a) how new strains of wheat are developed by researchers
(b) the new types of wheat that have been developed.
What big advantages does each new type provide for the growing of wheat in Canada?

2. Which products in your local supermarket and department store originate, completely or partially, from ranches? Construct a flow chart to show the stages in production and transportation starting at the ranch and ending at your supermarket and department store.

3. Before the modern development of chemicals to fertilize the soil and control insects and diseases, crop rotation was practised on many mixed farms. Explain carefully what is involved in rotation, using examples of the use of the land. Explain how these techniques helped to keep the soil fertile and reduce insects and disease damage.

4. Make a detailed study of one of Canada's major fruit-growing areas other than the Okanagan Valley. Your presentation should include
(a) a map and description of its location
(b) an explanation of its advantages for fruit growing
(c) the quantities of different kinds of fruit grown
(d) the packaging or processing of the fruit for the consumer.

5. Choose one of the following vegetables and explain
(a) how the soil is prepared
(b) when and how the seeds are planted
(c) how weeds, insects, and diseases are controlled
(d) any special care, such as irrigation
(e) the equipment and methods used to harvest the mature crop
(f) how the product is prepared for market
(g) how the vegetables are stored and transported.

potatoes turnips carrots lettuce cabbage peas
celery field tomatoes hot-house tomatoes beans

6 WATER

The Importance of Water

Water is an essential part of everything that lives on earth. It comprises about 70% of your body and forms the greatest portion of most living plants and animals. Without water, present life on earth would not exist.

WATER MAKES UP...

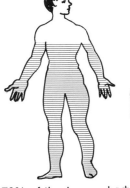

70% of the human body

Figure 6-1 Water is vital to life

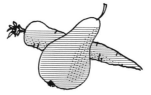

50-75% of meats

95% of edible fruits and vegetables

10-15% of dry grains and dried peas and beans

90% of milk

Every day you need to consume about 2 L of water, just to stay healthy. This water acts to flush out poisons, which would otherwise harm you. By dissolving harmful substances, water is able to carry them away and make your life possible. Since water can dissolve so many substances, it has been called the **universal solvent.**

Our daily life is centred on water. Each person in an average Canadian household, for example, uses 225 L of water every day. Few Canadians ever venture far from a source of water whether it is in a jug, a faucet, or in food.

Water is an essential ingredient in many activities, such as tourism, farming, steel making, and baking bread. It is also used as a basis for measurement of many other items, as you can see in Figure 6.2.

Figure 6-2 Water is important in measurement

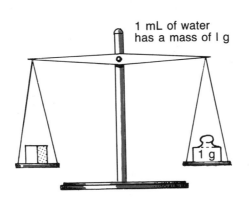

1 mL of water has a mass of l g

1 g

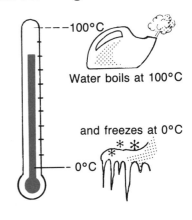

100°C

Water boils at 100°C

and freezes at 0°C

0°C

1. Most Canadians take water for granted. Write a paragraph about your daily life to show the ways in which you use water. Part of your answer should refer to the water you waste every day.
2. Assume that water was suddenly rationed (limited) to 10 L for each person every day in your local area. How would you be forced to change your way of life to survive? Be specific in your answer.

Water Supply and Use in Canada

Canada is a unique country in terms of water supply. It has a longer ocean shoreline and a greater area of fresh water than any other country in the world. We also have more fresh water per person than any other country.

With all of this water, some people have made the mistake of assuming that it is not important to manage the water properly. As a result, some of our water supplies are badly polluted. As well, there are some regions where there is not enough water to supply all the requirements of the people who live there. Since water is essential for so many purposes, we must be careful to manage our water so as to avoid waste and pollution as much as possible.

---

3. Using Photographs (a) to (f) in Figure 6.3 as a starting point, list as many uses of water as you can think of. You should have at least 20 examples.

Figure 6-3 Several uses for water

THE HYDROLOGIC CYCLE

All the fresh water in Canada originated from the oceans. The heating of the ocean water by the sun is the key process that starts the hydrologic cycle (water cycle) in motion. Eventually fresh water reaches the land to fill lakes, rivers, and other forms of surface water, as shown in Figure 6.4.

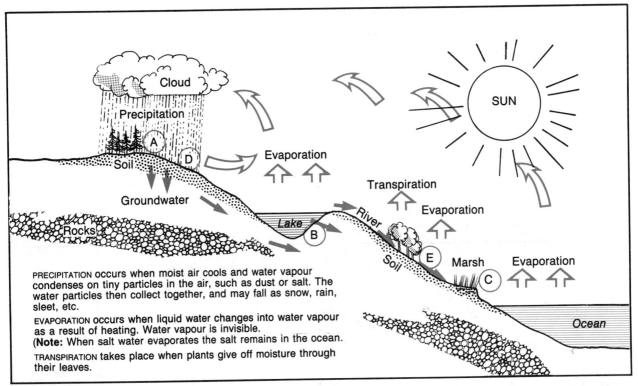

PRECIPITATION occurs when moist air cools and water vapour condenses on tiny particles in the air, such as dust or salt. The water particles then collect together, and may fall as snow, rain, sleet, etc.

EVAPORATION occurs when liquid water changes into water vapour as a result of heating. Water vapour is invisible.
(**Note:** When salt water evaporates the salt remains in the ocean.)

TRANSPIRATION takes place when plants give off moisture through their leaves.

Figure 6-4 The hydrologic cycle

4. (a) Examine Figure 6.4 carefully. Write a full description of how water moves and is changed in form during the hydrologic cycle. Start with the heating of the ocean.
   (b) Describe the place where you live in relation to the water cycle.

5. What would happen if the hydrologic cycle were broken because one of the steps you described in answer to Question 4 no longer occurred? Explain your answer fully.

6. Referring to Figure 6.4, write a description to outline the impact on the hydrologic cycle in each of the following events.
   (a) The chopping down of trees at A.
   (b) The draining of the lake at B and the marsh at C.
   (c) The building of a city with paved streets and large buildings at D.
   (d) The construction of a dam on the river at E.

THE IMPORTANCE OF PRECIPITATION

The amount of water that any area in Canada receives depends mainly on the amount of precipitation and river flow available. The amount of precipitation, however, can vary greatly from season to season in any location. Some Canadian regions experience very heavy precipitation in some seasons, and relatively little in others. The results of these variations mean that pipelines and large-scale dams must be used to supply some cities with water throughout the year.

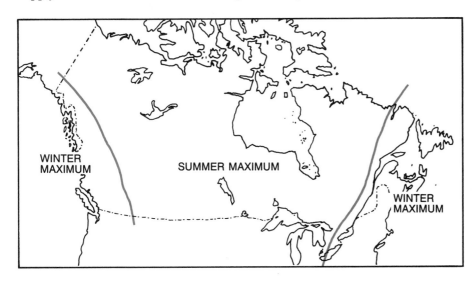

Figure 6-5 Seasons of peak precipitation

In Figure 6.5 Canada is divided into regions according to the season of peak precipitation. On the east coast, for example, the precipitation reaches a peak during the winter. This is also true on the west coast, although the west has a much higher total yearly precipitation.

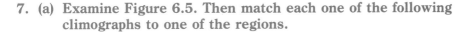

7. (a) Examine Figure 6.5. Then match each one of the following climographs to one of the regions.

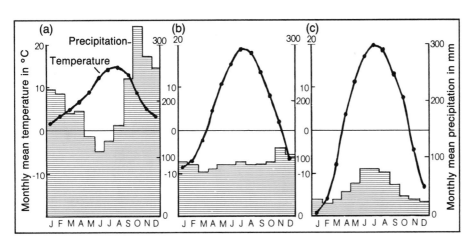

(b) Give reasons for your answers to **7 (a)** and describe the precipitation pattern for each station.

(c) Which of these stations has a precipitation pattern best suited for agriculture? Explain your answer fully.

(d) At which of these stations might you expect the heaviest winter snowfalls? Explain how you arrived at your answer.

8. (a) What seasonal pattern of precipitation do you have in your home area? Include in your answer the amount and type of precipitation. Refer to Figures 2.36 (page 51) and 6.5.

(b) What influence, if any, does this have on your activities throughout the year?

WATER RUN-OFF

Precipitation that falls to the ground and then flows into our streams, rivers, and lakes is considered to be **run-off.** Approximately 48% of all precipitation in southern Canada becomes run-off. These surface sources of water are the most vital ones in Canada.

Although the amount of precipitation an area receives has a strong influence on the run-off, there are other important factors. One of these is temperature. In the winter, for example, snowfall is the dominant form of precipitation in Canada. However, snow does not add greatly to the amount of run-off at the time it falls. Only when the temperature rises in the spring does the snow melt and substantially affect the run-off.

The snow cover in forests melts more gradually than in open areas such as the Prairies. This means that the run-off from forested areas causes less flooding than that from non-forested areas.

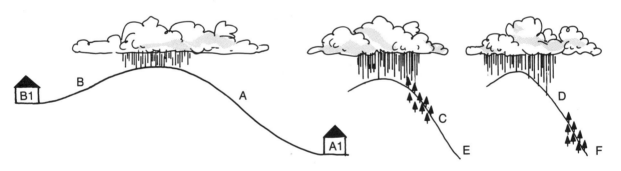

9. Imagine that you have to locate a city at either A_1 or B_1. Explain which site would be the most suitable. Consider the danger of possible floods. Explain the reasons for your answer fully, referring to areas A and B.

10. (a) Rank sites C, D, E, and F in order according to the danger of flooding there. Place the safest site first. Give reasons to explain your answer.

(b) What conclusions about flood prevention can you think of as a result of answering 10 (a)?

The Variation in Seasonal Run-Off

Precipitation, as you have discovered, varies from one season to another. In a similar way, run-off in Canadian rivers changes according to the season. This fact is illustrated in Figure 6.6.

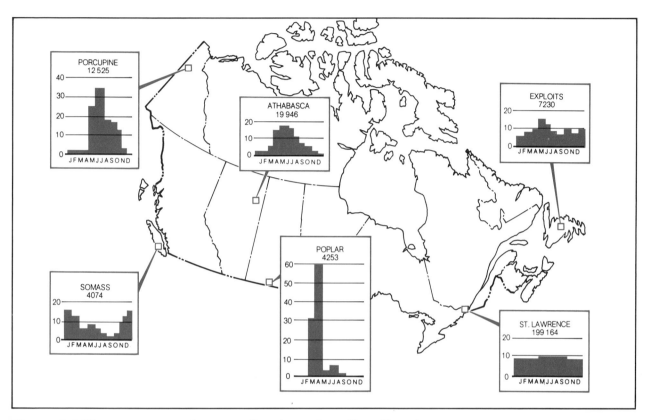

Figure 6-6 The average yearly water flow in various Canadian rivers
The graphs show the volume of flow each month as a percentage of the yearly flow. The number equals the annual average flow in thousands of cubic metres.

11. (a) Examine and describe the flow of each river shown in Figure 6.6. For your answer, construct a table similar to the one below, and then complete it. Refer to Figure 6.6 and earlier portions of this chapter to help with your answer.

NAME OF RIVER AND REGION	SEASON(S) OF MAXIMUM FLOW	SEASON(S) OF MINIMUM FLOW	CLIMATIC REASONS FOR PATTERN OF FLOW
	SAMPLE	ONLY	

(b) Which river appears to pose the greatest flood threat? Give reasons for your answer.

12. (a) One important factor for river transport is a reliable year-round flow of water.
    (i) Based on Figure 6.6, name the river on which ship transport would seem to be most difficult. Explain your answer.
    (ii) Which river appears to be most suitable for ship transport? Why?
    (b) What other factors would influence the suitability of a river for ship transport?
13. Explain the reasons for differences in water flow between the Porcupine and St. Lawrence Rivers. Use the climate map in your atlas as an aid.
14. What specific problems would a town experience if it was built along the banks of the Poplar River?

GROUNDWATER

Water that does not become run-off and instead sinks into the soil is called **groundwater.** Some groundwater flows into underground rivers and streams that might run through caverns or similar formations. A great deal of this groundwater, however, is found in **pores** (holes) in the rock and soil.

Groundwater forms an important source of water for Canada's population. Wells, for example, can tap groundwater for human uses. They provide an invaluable source of water, especially in the drier regions of British Columbia, the Prairies, and even the more humid parts of eastern Canada.

The **water table,** shown in Figure 6.8, is the level at which there is water available when a well is dug or drilled. If the water table reaches the surface of the ground, a lake, river, or other water feature occurs.

Figure 6-7 Water emerging from an underground cave

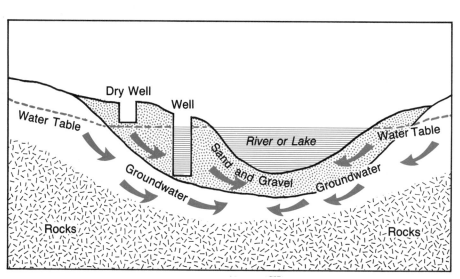

Figure 6-8 Groundwater and the water table

THE DRAINAGE BASIN OF A RIVER

All Canadian rivers share very similar overall patterns. Although they may flow through different landscapes, the various parts of the rivers are the same. The Fraser River (Figure 6.9) is an important Canadian river. Each of its parts is shown.

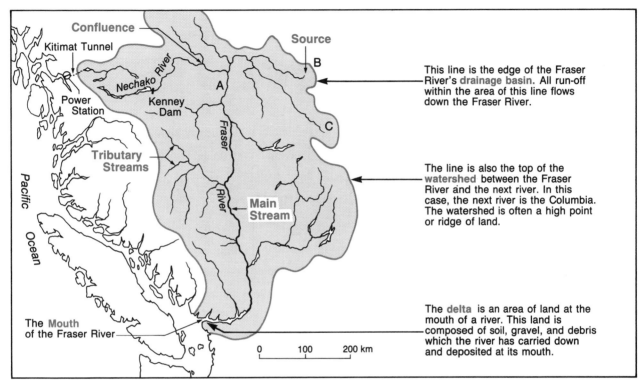

Confluence

Kitimat Tunnel

Source

B

This line is the edge of the Fraser River's **drainage basin.** All run-off within the area of this line flows down the Fraser River.

Nechako River

A

Power Station

Kenney Dam

C

The line is also the top of the **watershed** between the Fraser River and the next river. In this case, the next river is the Columbia. The watershed is often a high point or ridge of land.

Tributary Streams

Fraser River

Main Stream

Pacific Ocean

The **delta** is an area of land at the mouth of a river. This land is composed of soil, gravel, and debris which the river has carried down and deposited at its mouth.

The **Mouth** of the Fraser River

0   100   200 km

Figure 6-9 The drainage basin of the Fraser River *Take note of the Kenney Dam, which blocks the Nechako River. A lake has been formed behind the Kenney Dam. The water from this lake flows west to the Pacific Ocean. On the way, the water generates electricity for the aluminum refinery at Kitimat.*

15. (a) Trace the outline of the Fraser River into your notebook. Print arrows on each branch of the river to show the direction in which the water flows.
    (b) In which compass direction does the river flow between A and B and between A and C?
16. Explain in your own words the meaning of each of these terms, as illustrated in Figure 6.9.
    (a) mouth (b) tributary (c) confluence
17. (a) What would happen near the mouth of the Fraser River if there was an extremely heavy rainfall at A?
    (b) What else could occur near the source of the Fraser River that could help or hurt those people living downstream? Explain your answer fully.
    (c) If you were a government planner, would you make plans for the Fraser River on the basis of
    (i) each tributary separately?
    (ii) the entire river?
    (iii) other sections of the river based on your own divisions?
    Explain how you arrived at your answer.

Canada is divided into a series of major drainage basin areas. These are shown in Figure 6.10.

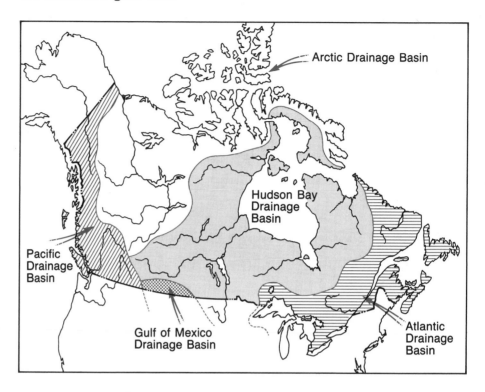

Figure 6-10 The major drainage basins of Canada

18. In what large drainage basin do most Canadians live? Refer to the population density map in your atlas.
19. (a) Into what ocean does most run-off in Canada flow eventually?
    (b) What problems does Canada's population have as a result of this situation?
20. (a) In what drainage basin do you live?
    (b) Describe the route water would take to reach the ocean after leaving your local area. Refer to your atlas for help on this one. (You may also need a local map.)
21. Assume you are a government water researcher. Your task is to discover where water flows after entering Lake Superior at Thunder Bay, and the uses made of it.
    Use a table such as the one shown below to list the major uses for water between Thunder Bay and the Atlantic Ocean. Explain where each of these water uses dominates. Use the thematic maps in your atlas to help.

MAIN WATER USES	MAJOR AREAS FOR THIS USE
1. 2. **SAMPLE**	**ONLY**

WATER SURPLUS: PROBLEMS AND SOLUTIONS

When water accumulates in an area more quickly than it can be carried away, there is a **water surplus.** This surplus may result from a brief but violent thunderstorm, melting snow, or long, hard rainstorms. The surplus may go unnoticed by local residents, or it may cause severe flood damage as in Figure 6.11. These major floods have become a source of great concern for Canadians in certain regions.

Figure 6-11 A farm truck crossing a flooded field in early spring

Most flooding occurs during the spring. At that time snow melts quickly as temperatures rise. The soil is still largely frozen, so it will absorb little water. As a result, the melt-water from the snow quickly fills up the rivers and overflows their banks. This floods the surrounding land.

A great deal of flooding in Canada occurs in unpopulated areas. When a flood does hit a settled area, however, the cost may be millions of dollars, in addition to the loss of human or animal life. Figure 6.12 indicates areas where serious flooding often occurs. Each river that floods causes some destruction to human settlement.

---

22. (a) Using your atlas map of Canada, list the major Canadian cities that are often affected by flooding.
    (b) Which rivers flood good agricultural land?
    (c) To which oceans do the flooding rivers eventually flow?
    (d) How would you describe the distribution of flooding across Canada?
23. Why do you think Nova Scotia and Prince Edward Island usually escape major flooding? Refer to detailed regional maps in your atlas.

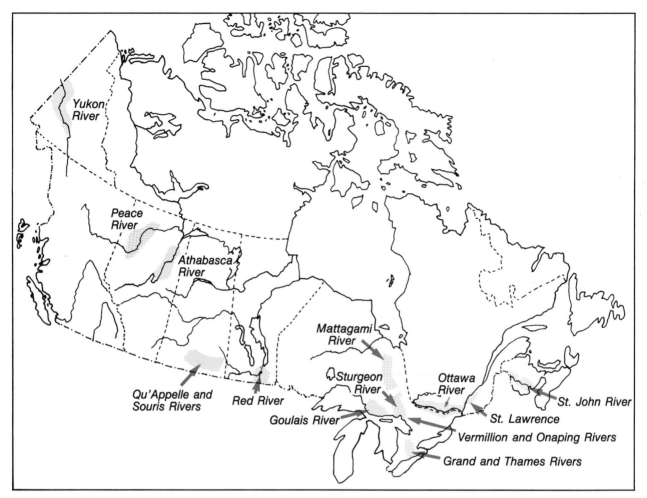

Figure 6-12 Areas frequently affected by spring flooding

24. **Which three effects of flooding do you consider to be most serious? Explain the reasons for your answers. (Flooding results in damages to road and rail transportation; human, farm, and wild animal life; agricultural crops; industries, housing, and pure water supplies.)**

The government of Canada has become increasingly concerned about flood damage. In order to try to reduce this damage, a number of **flood-risk maps** have been developed. Such maps show the zones that have been flooded in the past. These zones also risk damage in any future flood. Figure 6.13 shows such a flood-risk map for Carman, Manitoba.

To discourage building in the vulnerable "floodway zone," laws were established in 1984 prohibiting the building of government buildings there. No government grants or damage assistance will be given to anybody who builds there after June 1984. The area called "floodway fringe" can expect a flood once every 100 years, on average. It is permitted to build there as long as the buildings are floodproofed.

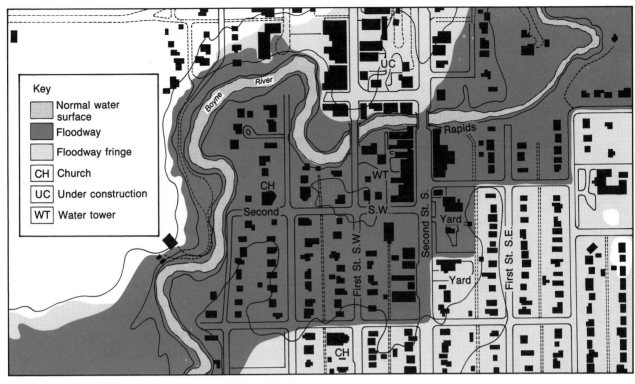

Key
Normal water surface
Floodway
Floodway fringe
CH Church
UC Under construction
WT Water tower

Boyne River
UC
Rapids
WT
CH
S.W.
Second
First St. S.W.
Second St. S.
Yard
First St. S.E.
Yard
CH

Figure 6-13 Flood-risk map for Carman, Manitoba

25. You are employed as a reporter for the Winnipeg Free Press, and are given the assignment that follows. Read over the special assignment sheet and write out the report marked with an asterisk (*).

SPECIAL ASSIGNMENT SHEET

Occupation	Newspaper reporter
Home Base:	Winnipeg Free Press
Special Assignment:	To report in detail about the flood damage in Carman, Manitoba.
Transportation:	Helicopter to the site of flooding in Carman.
Background Brief:	Figure 6.13. The floodway is completely covered with water.

*The Report to be filed before you leave includes
(1) half a page on the buildings, roads, and facilities that have been flooded
(2) the location and subjects for 15 photographs that you intend to take
(3) three types of people, each with a different occupation, that you intend to interview while you are in Carman.

Case Study: The Winnipeg Flood Program

Flat land is ideal for many types of agriculture, as we have seen. During periods of flooding, however, such land can be easily covered with water. Cities located on the Prairies, such as Winnipeg, are particularly open to flood damage.

Winnipeg has experienced a number of devastating floods. As a result, a flood program was introduced in the 1950s. Figure 6.14 indicates the main features of the plan. Although this program was referred to as ''folly'' (a foolish waste of money), it paid off in 1979 when it saved Winnipeg from what might have been its most damaging flood.

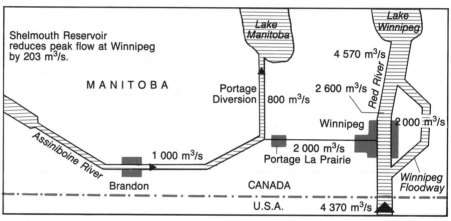

Figure 6-14 The Red-Assiniboine flood control system

NOTE: The figures and the width of each part of the system represent the maximum amount of water that can be held without flooding over the banks. (m³/s = cubic metres per second.)

26. Explain in your own words how the floodway works.
27. Why do you think people referred to the floodway plan as folly?
28. The Winnipeg floodway program cost $61 276 000. What costs would have been included in this sum?

WATER DEFICIT: PROBLEMS AND SOLUTIONS

Most areas of Canada have experienced a drought of some description. Droughts occur when there is a serious shortage of water in a certain area for farming and human activity. At times a drought might last only a few weeks. In some situations a drought could extend over a number of years. The effects of a drought, therefore, can vary significantly.

Water deficits include droughts and can occur in a regular pattern. In Figure 6.15 you can see those areas of Canada that, on the average, have a water deficit every year. Whenever the amount of water available for plant growth is too small, a water deficit exists.

29. (a) What regions of Canada experience the most serious water deficits?

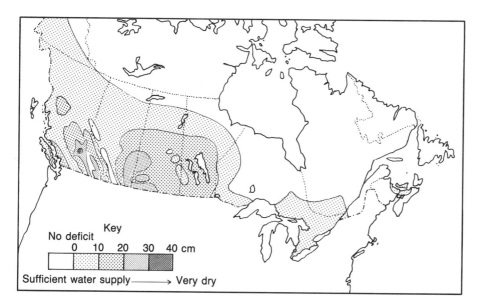

Figure 6-15 Water deficits in southern Canada

Key

No deficit

0   10   20   30   40 cm

Sufficient water supply ⟶ Very dry

(b) Considering that the general movement of air in Canada is from the west to the east, what contributes to the deficits in 29 (a)?

(c) What could be done in these regions to make farming profitable?

30. (a) Is there a water deficit in your local area? If so, how great is it?

(b) What indications have you seen during the late spring, summer, and early fall that this water deficiency exists in your local area?

31. In what areas of Canada are there no water deficits during an average year?

As technology improves, attempts can be made to overcome the effects of water deficits. Despite some undesirable side effects, dams are widely used to store water and control its flow. Water held in dam-created lakes can be drawn upon in times of deficit to help nearby areas. Water that flows into a river system during periods of high run-off can be held back by a dam until needed in a period of drought. As you can see, dams help control floods. They also lessen the damage of droughts.

32. What other uses, besides control of water flow, might the artificial lake of Figure 6.16 have?

33. What special conditions must exist in a river valley to allow a dam to be built and an artificial lake to be created? Explain your answer fully.

34. The dam shown in Figure 6.16 (a) and (b) works efficiently in the conditions shown.
Under what weather conditions would the dam not operate so well? Why?

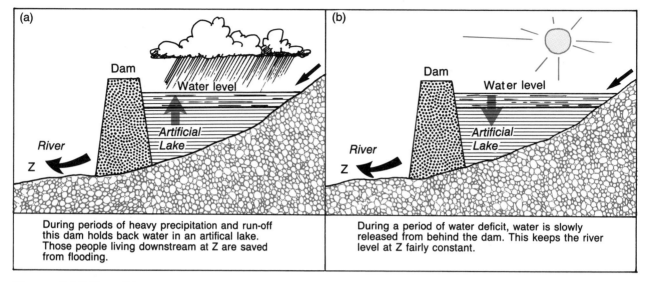

(a)

Dam

Water level

Artificial Lake

River

Z

During periods of heavy precipitation and run-off this dam holds back water in an artifical lake. Those people living downstream at Z are saved from flooding.

(b)

Dam

Water level

Artificial Lake

River

Z

During a period of water deficit, water is slowly released from behind the dam. This keeps the river level at Z fairly constant.

Figure 6-16 Benefits derived from building a dam

WATER MANAGEMENT

As our Canadian population grows, there is an increasing demand for reliable, clean supplies of water. At the same time there is an urgent need to plan the development of land along our rivers. If a river is not managed properly, there may not be enough water to supply everyone's needs.

In Ontario, for example, a series of conservation authorities have been set up. Each authority attempts to manage one or more rivers and their watersheds. Such management involves the building of dams, tree planting, and development of parks. In this way people can enjoy recreation in the river valleys, but flooding is controlled.

---

35. (a) Where are most conservation authorities located relative to Ontario's urban centres?
    (b) Explain why these locations are especially valuable in Ontario.
36. Choose one of the conservation authorities in Figure 6.17 and open your atlas to a map of southern Ontario. What specific cities would benefit from
    (i) flood control projects on the river?
    (ii) recreational areas along the river?

There have been numerous plans and actual projects across Canada that have been designed to manage our water resources. One plan, called NAWAPA, has been proposed to divert Canadian water south. Another water management scheme, which is presently underway, is the James Bay Project in Quebec. Both projects involve issues of concern to Canadians.

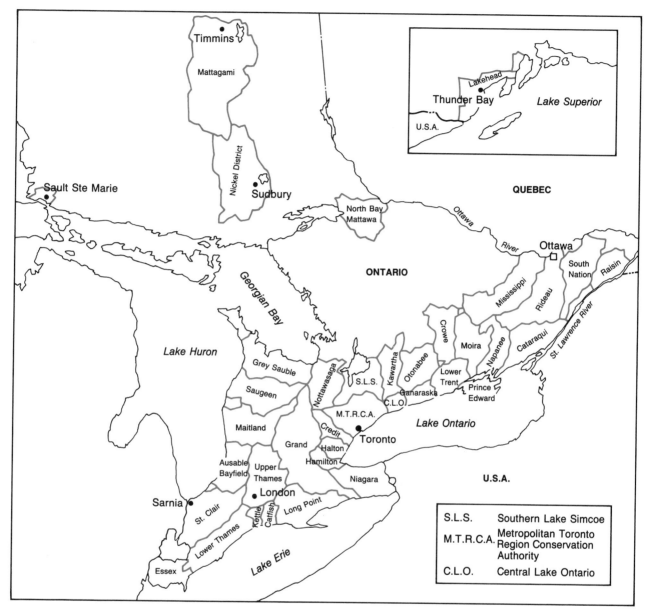

Water Transfer

As you have already discovered, water is often needed in regions where it is not naturally available in sufficient quantities. As a result, a number of schemes have been suggested to transfer water from regions of surplus to regions where it is needed. Some of these proposed water-transfer schemes have already been implemented. The Nechako River in British Columbia, for example, was partially diverted, as shown in Figure 6.9 on page 189. The diverted water was dammed to form a lake, which then flowed west to provide electricity for Kitimat's aluminum refinery. Most existing water-diversion (transfer) schemes have been developed to provide hydro-electric power.

Figure 6-17 The conservation authorities of Ontario

Case Study: The Nawapa Proposal

All of the existing water-transfer programs appear to be tiny in comparison to a proposal called NAWAPA (North American Water and Power Alliance). NAWAPA is a plan involving a giant series of dams, lakes, and canals designed to transfer water from north-flowing Canadian rivers to the United States. Water that flows to the Arctic Ocean is largely unused by Canada's population. NAWAPA proposes to divert some of this water south to the western and mid-western regions of the United States. Both of these regions are presently short of water and would like to buy Canadian water. Figure 6.18 shows the NAWAPA scheme.

Figure 6-18 The North American Water and Power Alliance (NAWAPA) proposal

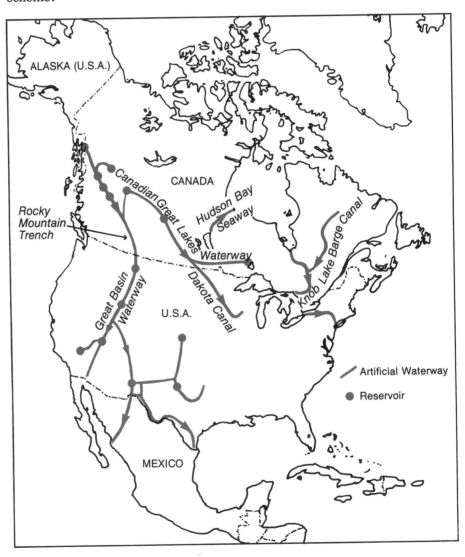

The NAWAPA scheme is referred to as a **continental resource program.** As a continental program, resources are transferred within North America to wherever they are needed.

37. **List three advantages and three disadvantages of the NAWAPA program for either Canada or the United States. Consider the impact of the dams, canals, and new lakes as well as the various uses the water would have in the United States.**

Case Study: *The James Bay Project*

Most of the province of Quebec is undeveloped wilderness. Sprinkled with thousands of lakes and numerous wild rivers, northern Quebec has been largely untouched by human progress. The forests and mineral resources of the region lie mostly undisturbed. In fact, the only means of transportation to most of northern Quebec is by canoe or airplane.

It is in one portion of this region that the giant James Bay Project was begun. This project, as originally designed, would affect about 20% of Quebec's total area. It would involve giant hydro-electric developments on at least five large rivers that flow into James Bay.

The first river to be developed was the La Grande River as illustrated in Figure 6.19. The La Grande project would involve the construction of numerous dams, spillways (artificial water ways), powerhouses, and artificial lakes. Water from other rivers would be diverted into the La Grande River to help in the production of hydro-electricity.

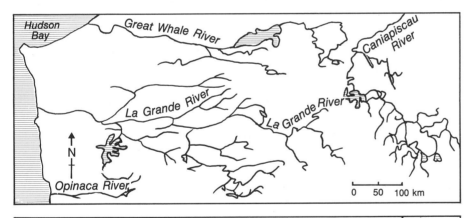

Figure 6-19 The La Grande River project

(a) Before commencement

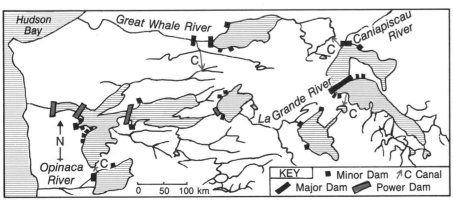

KEY ■ Minor Dam ⤴C Canal ◆ Major Dam ▬ Power Dam

(b) After completion

38. (a) Using your atlas as a guide, describe the location of the La Grande River.
    (b) How long is the La Grande River? In which direction does it flow?
    (c) Turn in your atlas to a map of your home province. Starting where you live, measure the number of kilometres you recorded in 38 (b). Which Canadian city could you reach in that number of kilometres? This answer will give you some idea of the size of this first step in the James Bay Project.
39. What three major changes shown in Figure 6.19 will the James Bay Project bring to the La Grande River?

Since the James Bay Project was launched in 1971, there has been a great deal of debate over its real value. Below is a summary of the discussion for and against the project.

In Support of the James Bay Project	In Opposition to the James Bay Project
• It will supply needed hydro-electricity to Quebec. • Money can be earned by selling power to the United States. • It will provide many thousands of jobs for Quebeckers. • The project will involve building numerous roads and airports as well as communication links. • New resources in the James Bay region such as iron ore and timber would be developed. • The Indians can fish and hunt elsewhere or try new ways to make their living. • There has been little study to show that the balance of nature will be upset. • Tourists will bring money to the area as they come to use the lakes and the rivers.	• With increasing conservation of energy, this power may not be needed. • Much electricity is lost in power lines when it travels long distances, such as to New York State. • Fewer jobs than first predicted will be provided. • Transportation and communication links could be provided more cheaply than with this huge project. • Some of these resources, such as timber, will be flooded. Over 6 000 native peoples will have their lives disrupted. • The artificial lakes will flood the hunting and fishing grounds of the Cree Indians. • It will upset the balance of nature, nesting grounds for birds, and the climate of the area with the artificial lakes. • Tourists would enjoy the natural landscape better. There is no indication that more tourists would come to the region.

40. (a) Examine the table above. Choose the two most important points in support of the James Bay Project. Explain the reasons for your answer.
    (b) Repeat 40(a) for the two most important reasons to oppose the James Bay Project.

200

41. Write a short paragraph to indicate whether each of the
    following people would support or oppose the James Bay
    Project, giving full reasons for the point of view of each:
    (i) an unemployed Quebec worker
    (ii) a Cree Indian
    (iii) an American tourist
    (iv) a Quebec manufacturer of heavy machinery
    (v) an environmentalist
42. The La Grande Project has now been fully completed. There
    are plans to proceed to the other rivers in the plan. Do you
    think this should be done? Write a page to explain the reasons
    for your answer.

Pollution

Besides providing an adequate supply of water for human activities, we
are faced with the additional problem of cleaning up our polluted rivers,
lakes, and oceans. The sources of these pollutants are illustrated in
Figure 6.20.

Some of our water has been so badly polluted that not only is it unfit
to drink, it is also unsafe to swim in. Many of the fish living in these
waters contain too many chemicals to be safe to eat.

**Figure 6-20 The major
sources of water
pollution**

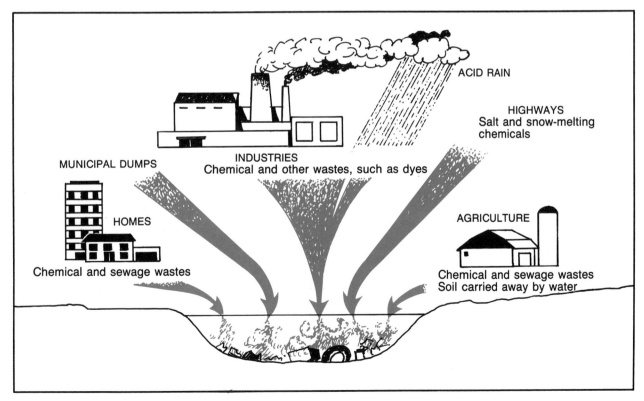

Figure 6-21 An area where swimming is prohibited

The Wabigoon River in Ontario has been closed for fishing because of high levels of mercury in the fish. This pollution has resulted from the wastes of the pulp and paper industry. Salmon have been successfully introduced into many rivers that flow into Lake Ontario. People are advised, however, not to eat too many of the salmon because of the high mercury content. Similar problems are developing elsewhere in Canada.

Another related problem is that of acid rain. Acid rain is common, especially in eastern Canada, where there is a great deal of industry. Figure 6.22 shows the sources and some of the effects of acid rain. Some of the acid rain results from pollutants in Canada and some is blown across the border from the United States.

Figure 6-22 The problem of acid rain

Solving the problem of acid rain is not easy. The first difficulty is that of finding the specific factories that pollute the air. Once this is done, there is the task of removing the pollutants from the smoke put out by the factory. Neither of the problems can be solved quickly, and in both cases the solutions are expensive to carry out.

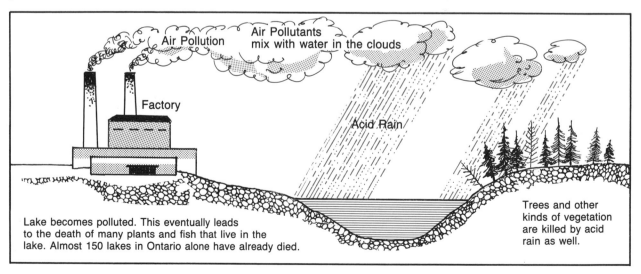

THE AGING OF OUR LAKES

All freshwater lakes in Canada age, or grow old. Like all living things, lakes have a certain life span after which they no longer exist. Figure 6.23 illustrates three stages in the aging process of a lake.

For most Canadian lakes this aging process takes thousands of years. With human interference such aging is greatly speeded up. Heavy water pollution, for example, can change a small, young lake fairly quickly into a middle-aged lake.

Lake Erie is a good example of a lake in its middle age. Many consider that Lake Erie is close to "dying." So much pollutants have been emptied into the lake that many fish have died. Certain portions of Lake Erie are choked with plant life. In addition there is up to 40 m of debris on the floor of Lake Erie. (The Cuyahoga River, which flows into Lake Erie at Cleveland, U.S.A., is so badly polluted that it has caught fire!) Many people believe that the quality of the water in Lake Erie is improving due to stricter pollution laws.

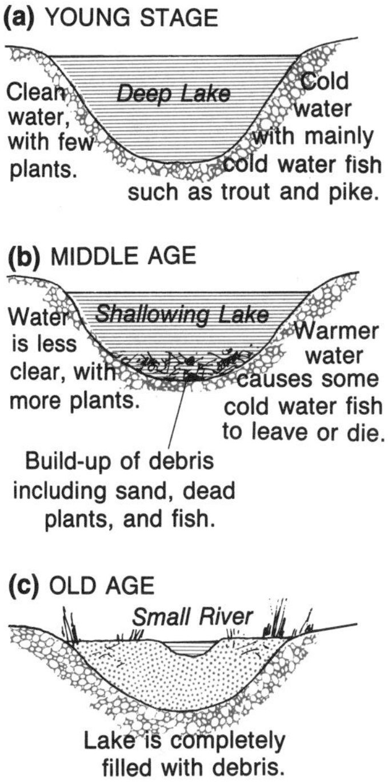

(a) YOUNG STAGE

Clean water, with few plants. Deep Lake. Cold water with mainly cold water fish such as trout and pike.

(b) MIDDLE AGE

Water is less clear, with more plants. Shallowing Lake. Warmer water causes some cold water fish to leave or die. Build-up of debris including sand, dead plants, and fish.

(c) OLD AGE

Small River. Lake is completely filled with debris.

Figure 6-23 How a Canadian lake ages

---

43. (a) List the bodies of water, streams, or rivers near your home that you suspect are badly polluted. What evidence do you have of their pollution?
    (b) Have you seen the sources of this pollution? If so, explain what the sources are.

44. You have had a chance to examine the various causes of water pollution. Using the thematic maps in your atlas, list those areas of Canada that you think would have the most serious water pollution problems. Give reasons for your answer.

45. Canadian water pollution problems cannot be solved without the help of the United States. Give two reasons to explain why this statement is true. Refer to atlas maps of North America to help in your answer.

46. Cleaning up our polluted waters involves a slow and very expensive program. In your opinion is this clean-up worthwhile? Give at least four good reasons to support your answer.

47. Imagine that you take a canoe trip down the Roger's River, starting at Hanville (see Figure 6.24). As you travel list the sources of probable water pollution that you will pass. Explain what actually causes the pollution. Make a table with the following headings and fill it in.

PLACE	CAUSE OF POLLUTION
SAMPLE	ONLY

48. What will happen to Clear Lake in the next ten years if water pollution is not controlled? Give reasons for your answer.

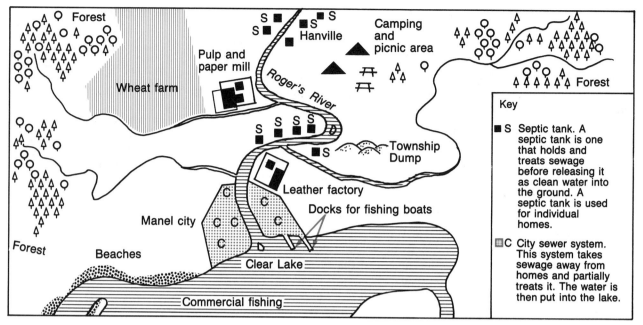

Figure 6-24 Roger's River

Key

■ S Septic tank. A septic tank is one that holds and treats sewage before releasing it as clean water into the ground. A septic tank is used for individual homes.

▨ C City sewer system. This system takes sewage away from homes and partially treats it. The water is then put into the lake.

There are, however, many damaging pollutants besides acid rain. All Canadians either directly or indirectly pollute our waters, or buy products whose manufacturers have polluted the water.

There are two major types of water pollution, as shown in the table below.

TYPE OF POLLUTION	EXAMPLES	EFFECT ON LAKES AND RIVERS
Degradable pollution. The substances can be broken down to become harmless when put into a body of water. Sufficient time must be given for this to happen.	Household sewage, animal waste, fertilizer from farms.	Oxygen is used to break down these substances. This process robs fish and plant life in the water of needed oxygen and eventually they may die. This is called eutrophication.
Non-degradable pollution. The substances that enter the water do not break down.	Mercury, salt from highways, many solid materials, such as plastics.	These may kill life in lakes and may make water unfit for swimming, drinking, or washing.

CONCLUSION

In this chapter we have examined Canada's exceptional fresh water resources. Although Canada has so much water, there is considerable controversy about how this water should be used. The decisions that Canadians face about their water resource development are major. If the right decisions are made, Canada can profit greatly from its water. It is important, therefore, for the Canadian public to be aware of these decisions and to be determined to make the wisest choices. Careful management by each Canadian can mean that clean water will be a valuable gift to future generations.

Universal solvent   Source   Delta
Hydrologic cycle   Mouth   Water surplus
Surface water   Tributary streams   Flood-risk maps
Run-off   Confluence   Droughts
Groundwater   Main stream   Water deficits
Pores   Drainage basin   Continental resource program
Water table   Watershed   Eutrophication

1. (a) Obtain a map of your local watershed from your teacher. Label the rivers, streams, and lakes as well as the urban centres in that watershed.
   (b) Put a star beside those places that might pollute the water.
   (c) In which areas of the watershed is there recreation, farming, or industry? Mark these on your map, using symbols and a key.
   (d) What dams are there in your local watershed? Mark them on your map.
   (e) Find out about the historic development of your local watershed. Write an account, illustrated with examples and maps where possible, to describe its history.
2. Refer to the end of Chapter 5 and the instructions given there on library research. Select one of the following river development projects for study:
   - The Nelson River
   - The Columbia River
   - The Peace River
   - Churchill Falls
   For your selected development project, answer the following questions.
   (a) Describe the location of the river. Use your atlas and its index to help you.
   (b) Draw a well-labelled map of the project.
   (c) From reference books you have found in the library, describe
      - when the development began
      - its cost
      - the area it covers
      - the purpose
      - criticisms of the project.
3. Examine a water resources map of Canada in your atlas.
   (a) Locate those areas of Canada where bodies of water form the boundary with the U.S.A. List them in order from west to east.

(b) List the advantages and disadvantages of water as an international boundary. Consider the following issues in your answer: pollution, fishing, smuggling, water transportation, and recreation.

4. Work on the following project in a group of three or four. Make a complete study of water in your local area and present your findings in a form suitable for mounting on a wall. Your project should include

(a) where your water comes from and how it is transported to your area

(b) where and how it is purified

(c) how it reaches the users after it is purified

(d) the various types of users in your area

(e) water pollution treatment facilities

(f) any special flooding or drought problems that you have experienced in the last three years, and their effects.

Include information about flood-control systems if there are any.

Illustrate your project with clear maps, diagrams, and photographs where possible.

7 ENERGY AND TRANSPORTATION

Everything you do requires energy. Standing up, sitting down, driving a car, or turning on a light — all of these activities depend on the use of energy. The following quiz has been designed to show the importance of energy and the degree to which you, individually, manage it.

Answer the following questions as honestly as you can.

QUIZ?

1. If you had a chance would you buy
   (a) a sports car?
   (b) an 8-cylinder car with air conditioning?
   (c) a compact 4-cylinder car?
   (d) a van?
2. On the highway, do you and your family normally drive at
   (a) 88-92 km/h?   (b) 70-87 km/h?   (c) more than 92 km/h?
3. Have you discussed the rising cost of energy in your home in the last six months?
   (a) Yes   (b) No
4. When you wash do you normally
   (a) shower?   (b) bath?
5. Does the washing machine that cleans your clothes recycle wash water for a second load?
   (a) Yes   (b) No   (c) Do not know
6. To get a drink of water do you let the water run to get cold?
   (a) Yes   (b) No
7. When you leave a room for more than five minutes, do you ordinarily leave the lights on?
   (a) Yes   (b) No
8. Which of the following appliances does your household regularly use?
   (a) electric can opener   (c) blow dryer for your hair
   (b) garage heater   (d) floodlights
9. What temperature is your home generally kept at during the winter?
   (a) 20°C   (b) under 20°C   (c) over 20°C
10. Evaluate this statement as true or false:
    Since you are only one person, you cannot really do anything to help cut down energy use in Canada.

Turn to page 237 to check your personal energy rating.

Energy Use

If you look at Figure 7.1, you will notice that much more energy is being used by each person today than at any time in the past. Our modern way of life depends on this high level of consumption.

1. Look at Figure 7.1 to help you answer these questions.
   (a) How many kilojoules of energy were used by the hunter each day?

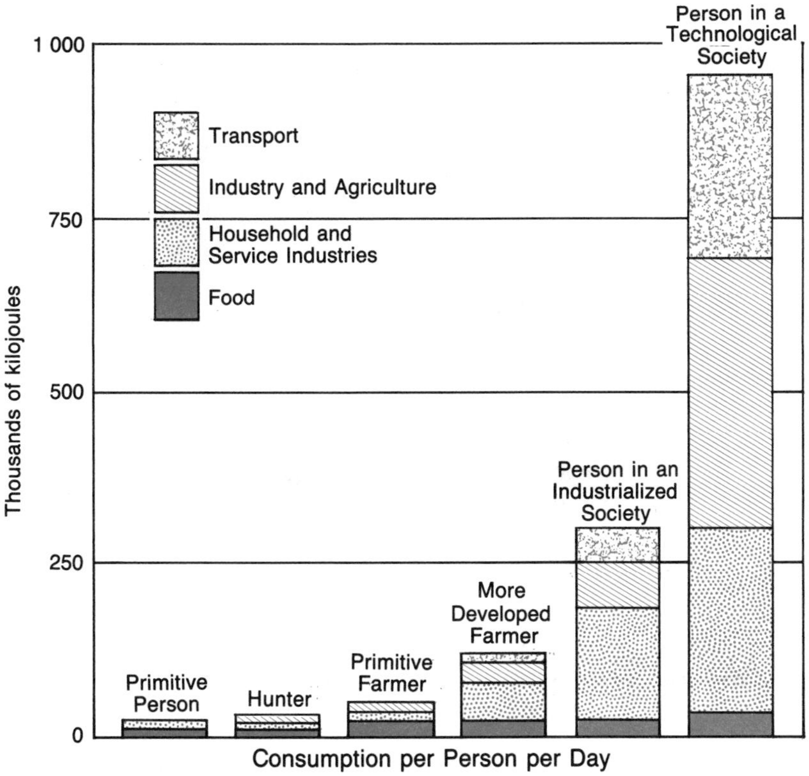

Figure 7-1 How energy use has increased

Thousands of kilojoules — Consumption per Person per Day

Legend:
- Transport
- Industry and Agriculture
- Household and Service Industries
- Food

Categories shown (left to right): Primitive Person, Hunter, Primitive Farmer, More Developed Farmer, Person in an Industrialized Society, Person in a Technological Society

(b) Apart from the food that the hunter ate, energy was used to run his house. What kinds of activities might have been involved in this?

2. How is energy used in a technological society? Look at the last column in Figure 7.1 to help you.

3. Why do you think that the amount of food consumed has not increased significantly?

4. What might happen to your way of life if the energy available to each Canadian was cut to one half of the present supply?

Canada uses more energy per person than any other country in the world. Factors that contribute to this include our cold climate, great distances to travel, and our high standard of living. Canada depends on six major sources to supply its energy needs.

SOURCE OF ENERGY	PERCENT OF TOTAL CANADIAN ENERGY NEEDS
Oil*	34
Hydro-electricity	25
Natural gas	19
Coal	12
Nuclear energy	6
Biomass (wood, etc.)	4

* This includes oil products such as gasoline.

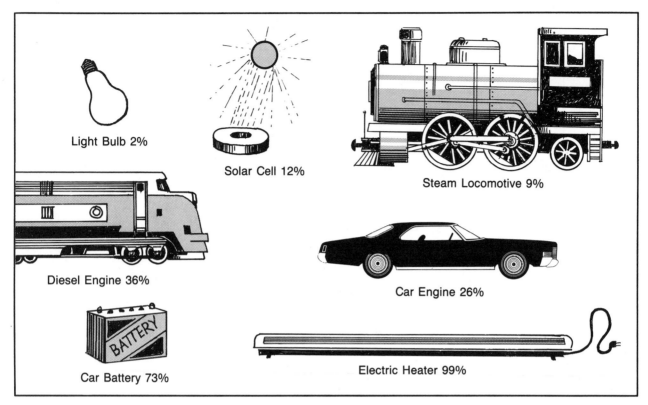

Figure 7-2 The efficiency of certain equipment

Light Bulb 2%

Solar Cell 12%

Steam Locomotive 9%

Diesel Engine 36%

Car Engine 26%

Car Battery 73%

Electric Heater 99%

ENERGY EFFICIENCY

Efficiency means the percentage of energy that goes into a device and actually comes out for the intended purpose. Have you ever burned your hand trying to unscrew a light bulb? It felt hot because only 2% of the electricity going into the bulb produced light; 98% produced heat. Such a bulb is very inefficient.

---

5. Rank the devices shown in Figure 7.2 according to their efficiency, from highest to lowest. Also note, in brackets, the percentage of the energy wasted for each device.

Renewable and Non-Renewable Energy

There are two basic types of energy in use today. The first type is non-renewable energy. Coal, for example, is non-renewable because once it is burned it is gone forever. Coal does not replace itself. It can only be used once.

Renewable energy sources are ones that are constantly replenished. Solar energy, for example, is a renewable type of energy. The sun can heat a house one day and then can heat it every day afterwards. Solar energy is not used up as coal would be.

6. Look at the list of our six major sources of energy on page 209.
   (a) Are most of Canada's energy sources today renewable or non-renewable?
   (b) What does this eventually mean for Canada?
   (c) If we want to plan properly for the future, what must we do about our sources of energy?

OIL

Canada's most important source of energy today is oil. Oil is not only very important for supplying our energy needs; refined oil has a wide range of other uses.

The gasoline that you put in your car did not come out of the ground in that form. When oil is mined, it is in the form of petroleum (crude or unrefined oil) and is thick, black, and tarry. When petroleum is refined, it is changed into hundreds of forms which you use in your daily life.

Petroleum is changed into useful products through two processes called distillation and cracking. Distillation breaks down petroleum into various types of substances; each one of these substances is refined again through cracking. The product that emerges after cracking is one we can use.

Figure 7-3 Products of petroleum refining

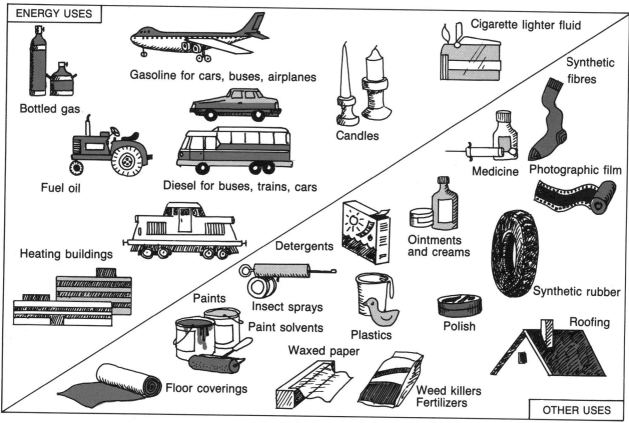

ENERGY USES

Bottled gas
Gasoline for cars, buses, airplanes
Fuel oil
Diesel for buses, trains, cars
Heating buildings

Candles

Cigarette lighter fluid
Synthetic fibres
Medicine
Photographic film
Synthetic rubber
Roofing

Detergents
Ointments and creams
Paints
Insect sprays
Paint solvents
Plastics
Polish
Floor coverings
Waxed paper
Weed killers Fertilizers

OTHER USES

Figure 7-4 The petroleum refining process *Powdered catalyst is passed into the reactor where the crude oil is cracked. The cracked oil then passes into a tall fractionating column where distillation separates the lighter and heavier parts of the oil. Hot air blows the catalyst back to the regenerator for re-use.*

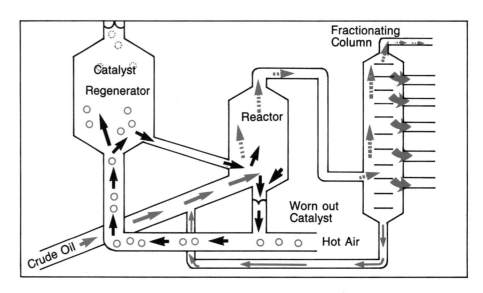

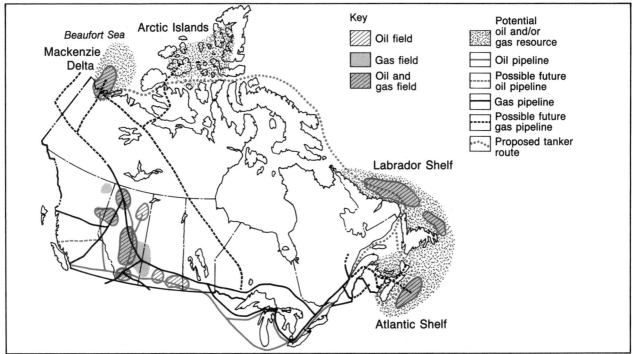

Figure 7-5 Canada's oil and gas resources

7. As you can see in Figure 7.5, Canada has extensive oil and natural gas deposits. Describe the location of the following:
   (a) areas of current production
   (b) areas that might produce these fuels
8. (a) Where are most of the potential (possible) but undeveloped deposits?
   (b) In light of your answer to 7 (a), what major problems does Canada face in attempting to develop its oil and natural gas deposits?

Petroleum is not easy to find. Even though Canada had the first commercial oil well in the world—near Petrolia, Ontario—it was many years before much oil was produced.

As early as 1884 a thick, black, tarry substance was discovered oozing to the surface of the earth in the Waterton Lakes region of Alberta. For some it was simply a nuisance, since it fouled the drinking water of their horses or mules. Others passed it by, considering it next to useless.

Many years later it became clear that this material that oozed up through cracks in the ground was petroleum and that it was exceedingly valuable. Today it has become so valuable that it is called "black gold." This name is appropriate for a substance that has become part of the foundation of our whole way of life.

The first problem that oil companies face is locating the petroleum. Most oil is not found oozing to the surface, as in the story above. Rather, crude oil is often buried quite deep under the surface of the earth.

There are many methods of discovering where oil is. Air photos, gravity meters, and magnetometers can all be used in helping to locate underground oil.

In Canada a common type of equipment used is the **seismograph.** The seismograph crews search for oil by setting off miniature earthquakes. Tiny explosions set off by these crews send vibrations or shock waves through the ground. (Shock waves in the ground are similar to what happens when you knock the side of a bowl of jello.) These shock waves bounce off underground layers of rock and return to the surface of the earth. At the surface there are a number of geophones that pick up the shock waves. A seismograph in a truck nearby records each shock wave. Once this is done a geologist reads the records and can tell if there is oil below.

Exploring for oil costs the oil companies many millions of dollars every year. Once discovered the oil must be extracted. To do this, wells are drilled down to the layer of rock that contains oil. Usually, many

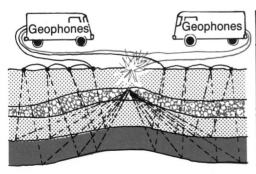

(a) **An explosion sends vibrations through the ground. The waves bounce off the layers of rock.**

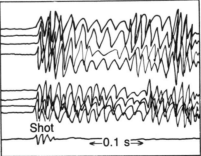

(b) **Geophones sense the vibrations and seismometers record them on graphs.**

Figure 7-6 Exploring for oil

wells are drilled before oil is found in large enough quantities to be pumped out. Before Imperial Oil Ltd. struck oil at Leduc, Alberta, in 1947, it had drilled 133 exploration (wildcat) wells.

Oil does not lie in a great pool below the earth's surface. It is actually found in the pores (holes) in rock. This porous rock is not unlike a sponge with different size holes in it. If you wanted to remove water from a sponge, you could use a straw to suck out the water. An oil well does basically the same thing. The shaft of the oil well links the oil-bearing rock to the surface and draws the petroleum upward. The pressure of the surrounding rocks and water help to push the oil up to the surface.

Natural gas is usually found with crude oil. Since it is lighter than oil, it is found above the oil deposits. At one time natural gas was burnt off as waste. In some countries this still occurs. In Canada, however, we have found out how valuable it is for such uses as heating homes and cooking.

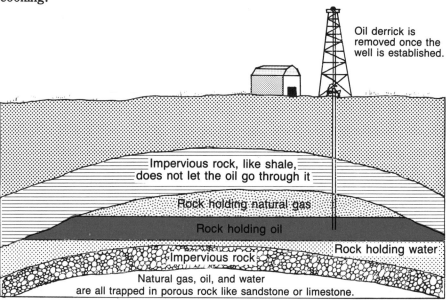

Figure 7-7 How oil and natural gas may be trapped in rocks beneath the surface

Offshore Drilling

As the oil companies explored for oil, they found some oil deposits in places that were difficult to reach. One of these places is below the ocean on what are called **continental shelves.** The Canadian continental shelf is a relatively shallow shelf of land that extends out from the land under the ocean.

Offshore drilling involves great costs such as those involved in setting up the platforms shown in Figure 7.8. Anchored to the floor of the ocean by huge chains, these drilling platforms are like floating islands.

The drilling crews must be extremely careful to avoid spilling oil into the water because this can kill fish and other sea life as well as destroy beaches.

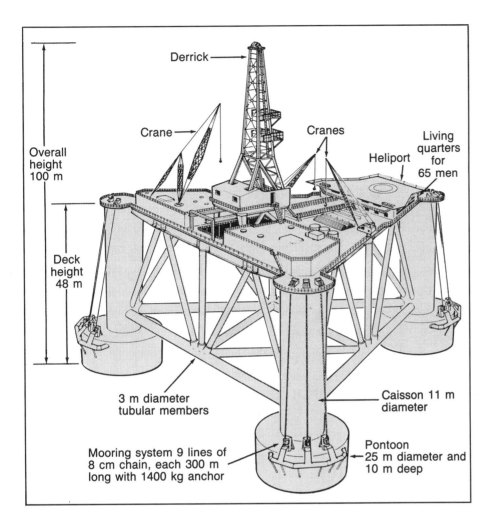

Derrick

Crane

Cranes

Heliport

Living quarters for 65 men

Overall height 100 m

Deck height 48 m

3 m diameter tubular members

Caisson 11 m diameter

Mooring system 9 lines of 8 cm chain, each 300 m long with 1400 kg anchor

Pontoon 25 m diameter and 10 m deep

Figure 7-8 An offshore drilling rig

Drilling in the Arctic

Beneath the frozen surface of Canada's Arctic lie great quantities of oil and natural gas. Some estimates indicate that beneath the Beaufort Sea lie more oil and gas reserves than exist in the rest of the world.

Drilling for oil in the Arctic is surrounded with difficulties. Transportation of workers and equipment into the Arctic for drilling, for example, is very expensive. Lack of roads and the fact that the Arctic Ocean is frozen most of the year add to these transportation problems. When oil is discovered in the Arctic, as it has been, there are major problems in moving it to the market.

Another difficulty centres on permafrost, which is permanently frozen ground. For most of the year the permafrost is frozen and therefore hard. During the brief Arctic summer, a thin layer of grass and vegetation protects the ground from significant thawing. During oil exploration, however, this vegetation is removed and the permafrost thaws. A small hole can quickly become a large gully, once the vegetation is removed.

Figure 7-9 Problems
using machinery in the
north

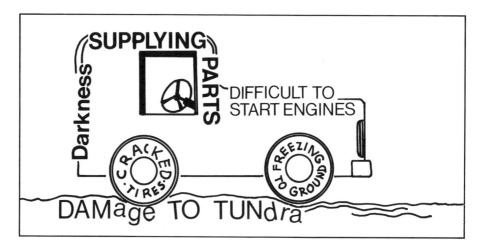

Clearly, permafrost presents a problem for the construction of roads or pipelines in the high Arctic. Nevertheless, the question remains: How can oil and natural gas be transported from the Arctic to the southern markets?

A pipeline is the most frequently discussed answer to this problem. The arguments for and against a pipeline through the Arctic are illustrated in Figure 7.10.

**Figure 7-10 Arguments
for and against an
Arctic pipeline**

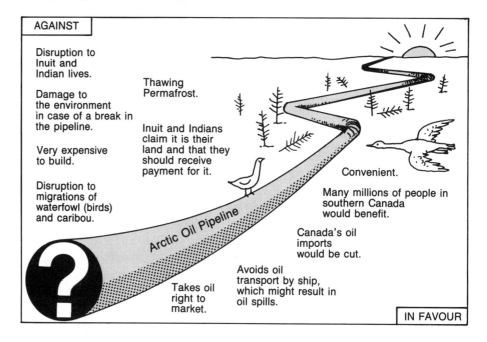

9. Examine both sets of arguments and select the point of view that appears to be most reasonable. Write a half page to explain the reasons for your opinion.
10. (a) What might cause a break in an oil pipeline in the Arctic?
    (b) What specific damage would such a break cause?

Oil and natural gas are moved from western Canada in pipelines, as shown in Figure 7.11, to the markets in eastern Canada. Pipelines are usually buried underground to protect against damage. The high cost of construction of these pipelines, however, limits the number that can be built. (When oil moves through a pipeline, it travels along about as quickly as you can walk.)

Figure 7-11 An oil pipeline under construction

11. Examine Figure 7.5 with its pattern of oil and natural gas pipelines in Canada. Turn in your atlas to a map of Canada's industry.
    (a) Describe which specific areas of Canada are joined by the pipelines.
    (b) What reasons can you suggest for this pattern of pipelines?
12. Turn in your atlas to a map of Canada's physical features. What specific problems did the builders of Canada's pipelines have to overcome?
13. What problems might a work crew encounter in trying to repair a break in an oil or natural gas pipeline?

Canada, at present, does not produce enough energy for its own uses. As a result, it imports large quantities of oil to the east coast. The oil is brought to the Maritimes by oil tankers from many countries, including Mexico and Venezuela.

Once petroleum has been refined, the petroleum products are shipped by train or truck to their destinations. Both the trucks and train cars are especially built to handle the different types of products from the refineries.

Figure 7-12 Canada's oil situation *Canada exports oil from the western provinces and imports oil to the east. Our oil imports now are greater than our exports.*

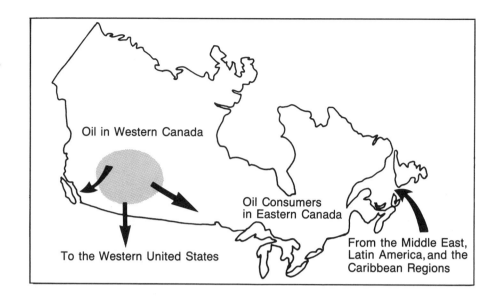

Oil in Western Canada

Oil Consumers in Eastern Canada

To the Western United States

From the Middle East, Latin America, and the Caribbean Regions

14. **Canada continues to export oil to the United States despite the fact that we must import oil for our own needs.**
    (a) **Why do you think Canada exports oil to the United States?**
    (b) **Should Canada continue to export oil to the U.S.A.? Give reasons for your answer.**

ALBERTA

Athabasca Oil Sands

Peace River

Fort McMurray

Cold Lake

Edmonton

100 km

Calgary

Figure 7-13 The nature and location of the oil sands *Oil sands are black and sticky. When frozen, the tar sands are harder than concrete.*

The Oil Sands

The statistics you studied earlier in this chapter indicate that oil is expensive. Oil is also difficult to obtain. These two conditions have led Canadians to exploration and development of the **oil sands** in western Canada. The oil sands are a sandy material that is saturated (completely filled) with oil. The appearance of the oil sands is not unlike that of asphalt used in paving roads.

The deposits of these vast oil sands are found under thousands of square kilometres of land, mainly in Alberta. Although these deposits are large, the problem is that they are buried underground — in some cases below many layers of rock and soil.

Chemical problems are another major obstacle to obtaining oil from the oil sands. It took many years of testing before the oil sands actually yielded oil that could be used in cars or for heating homes. Today Canada has some of the most advanced technology in the world for separating oil from the oil sands.

The process of extracting oil from the oil sands is illustrated in Figure 7.14. The size of the operation is large and involves many billions of dollars. The project shown in Figure 7.14 is one of the most expensive single developments in the history of Canada.

15. **In your own words explain the SUNCOR operation.**

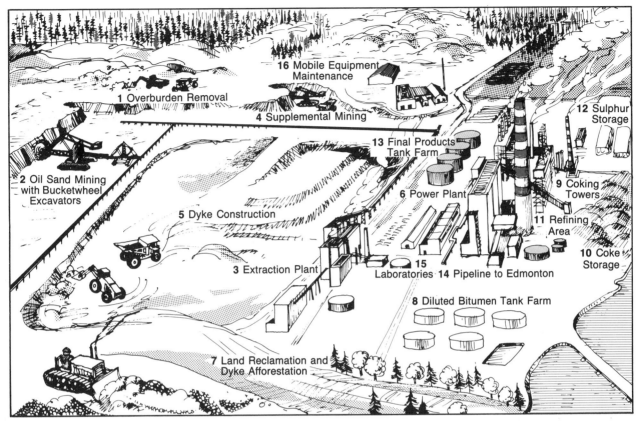

1 Overburden Removal
4 Supplemental Mining
16 Mobile Equipment Maintenance
13 Final Products Tank Farm
12 Sulphur Storage
2 Oil Sand Mining with Bucketwheel Excavators
6 Power Plant
9 Coking Towers
5 Dyke Construction
11 Refining Area
3 Extraction Plant
15 Laboratories 14 Pipeline to Edmonton
10 Coke Storage
8 Diluted Bitumen Tank Farm
7 Land Reclamation and Dyke Afforestation

The year-round mining of the oil sands requires a large labour force. To attract and accommodate workers, the city of Fort McMurray has been developed and modernized. New houses, schools, stores, and other facilities have been built to serve the new workers entering the city. Fort McMurray is an example of a **boom town** on Canada's northern frontier, since the settlement has grown so quickly.

Figure 7-14 The mining and extraction of oil from the oil sands *The SUNCOR project in Athabasca.*

Figure 7-15 The Syncrude operation — extraction and upgrading

ELECTRICITY

Almost everyone in Canada uses electricity every day. If the flow of electricity is interrupted for some reason, our lives are thrown into confusion. As Canadians we depend on electricity in so many ways.

16. List 25 items in your home that use electricity.
17. Assume that your local area lost its supply of electricity for 12 hours, starting at 20:00 h on a weekday. In what specific ways would you be affected?

The Cost of Electricity to You and Your Family

When you pay for electricity, you do so according to the amount of energy you use measured in **megajoules** (MJ). If a 100 W bulb is turned on for 10 000 s, the amount of electricity used is 1 MJ.

The difference in cost of operating various electrical devices is shown in the following table. Although there may be local variations, the price of electricity is assumed to be 1.7¢/MJ.

THE COST OF VARIOUS ELECTRICAL APPLIANCES

ELECTRICAL DEVICE	AVERAGE USE OF ENERGY PER YEAR (MJ)	COST PER MJ (1982)	TOTAL COST PER YEAR ($)
Clock	61.2 X	1.17¢ =	$ 0.72
Clothes dryer	4320 X	1.17¢ =	50.54
Frying pan	864 X	1.17¢ =	10.11
Hair dryer	54 X	1.17¢ =	0.63
Stove range	5580 X	1.17¢ =	
Shaver	2.16 X	1.17¢ =	
Colour television	1944 X	1.17¢ =	
Toaster	180 X	1.17¢ =	
Vacuum cleaner	162 X	1.17¢ =	
Light bulbs	6732 X	1.17¢ =	

NOTE: The expression MJ means megajoule. The figures in the table are Canadian averages; regional variations will alter them somewhat.

18. (a) In your notebook, complete the column of total costs per year.
    (b) Which device is the most expensive per year to operate?
    (c) On the average, which device is least expensive per year to operate?
19. From the list in the table, choose the devices that are used in your home. What is the total cost per year of all devices that are used?
20. In what specific ways could you cut down the electrical costs of your home?

How Electricity Is Produced

Until recently, most electric power in Canada was produced by running water. This is called **hydro-electricity,** and it now provides 75% of our needs. Because it depends upon water running downhill, it is renewable and has many advantages over other means of generating electric power.

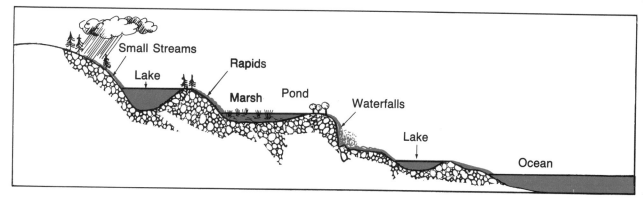

Figure 7-16 Profile of a typical Canadian river

HYDRO-ELECTRICITY

The steeper the slope that the water runs down, the more hydro-electricity can be produced. A waterfall is ideal in this situation since no dam need be built. A tunnel is built to take water from above the falls to a generator below the falls.

If a dam is built, it should be located in a valley with fast water or rapids upstream. When the dam is constructed, an artificial lake fills up behind. In effect, the dam creates an artificial waterfall that can be used to produce hydro-electricity.

Figure 7-17 A hydro-electric dam

21. Refer to Figure 7.16. Choose the best location for a hydro-electric development with no dam but with a tunnel. Explain your choice.
22. Where should you locate a dam and hydro-electric power station? What advantages does your location have over other places on the river?
23. Describe the least suitable location for any type of hydro-electric development. Explain your reasoning.

Hydro-electric power developments such as the ones cited earlier are considered to be clean. They do not produce significant water or air pollution. A number of effects of dam-building, however, can be discovered from the following diagram.

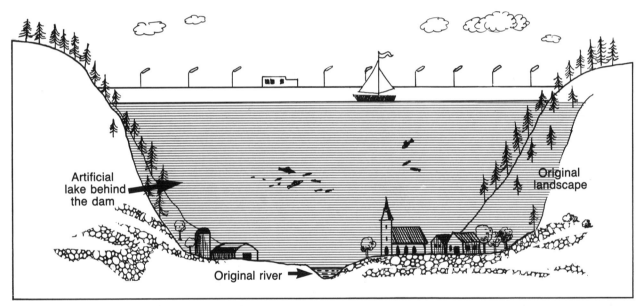

Figure 7-18 The impact of an artificial lake on the original landscape of a valley

24. List five effects that could result from the creation of an artificial lake behind a hydro-electric power dam.
25. In your atlas refer to the map of Canada that shows the locations of electrical plants. In what major watershed are most hydro-electric power plants found? Explain the reasons for this.
26. Which major river has the least hydro-electric development on it? What reasons can you suggest for this?
27. Describe and give reasons for the location of Canada's natural-gas electrical plants. Refer to Figure 7.5.
28. All electrical power plants are located close to water. Explain what water might be used for in each type of plant.

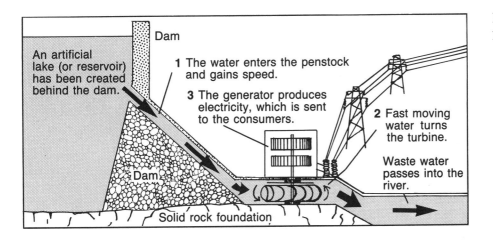

Figure 7-19 How water produces electricity

Dam

1 The water enters the penstock and gains speed.

3 The generator produces electricity, which is sent to the consumers.

2 Fast moving water turns the turbine.

An artificial lake (or reservoir) has been created behind the dam.

Waste water passes into the river.

Dam

Solid rock foundation

THERMAL ELECTRICITY

With **thermal electricity,** steam rather than water turns the turbine to produce electric power. The steam is produced when water is heated in a boiler, as shown in Figure 7.20. Fuels commonly used include coal, natural gas, and oil.

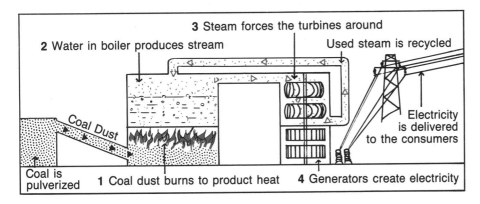

Figure 7-20 Thermal electricity production

3 Steam forces the turbines around

2 Water in boiler produces stream

Used steam is recycled

Electricity is delivered to the consumers

Coal Dust

Coal is pulverized

1 Coal dust burns to product heat

4 Generators create electricity

NUCLEAR ELECTRICITY

In principle, nuclear power stations are similar to thermal electrical plants. The heat that produces the steam in this case comes from a **nuclear reaction** of the uranium fuel, as you can see in Figure 7.21. When a nuclear reaction takes place, atoms are split and heat is generated. Only a small amount of uranium is necessary to produce a nuclear reaction.

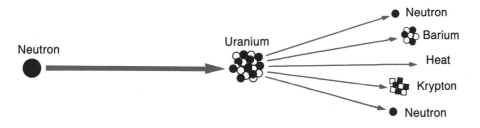

Neutron

Uranium

Neutron

Barium

Heat

Krypton

Neutron

Figure 7-21 A nuclear reaction *A neutron "bombards" a uranium atom. The atom is split into many different parts and this process releases heat.*

In Canada, the CANDU reactor is used to produce nuclear electricity. The CANDU reactor was developed in Canada and is one of the most advanced in the world. Although the reactor has been very safe, there is concern about the escape of radioactivity.

When uranium undergoes a nuclear reaction, radioactivity and heat are produced. This radioactivity is dangerous to human life and can kill. Numerous safety features have been built into the CANDU reactor to reduce the possibility of accident.

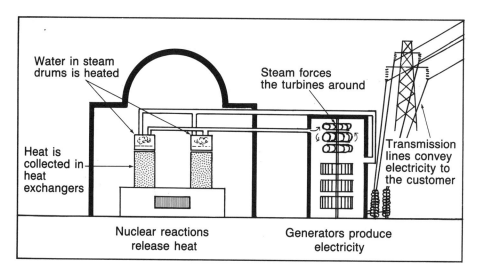

Figure 7-22 How nuclear electricity is produced

Peak Periods of Demand

Whenever you turn on a light switch, you expect the light to appear immediately. You do not want to wait a few minutes until light appears. This one example explains the unique character of electricity; it must be used as soon as it is produced. Electricity cannot be stored on any large scale. This fact poses many problems for the power suppliers. They must be able to generate electricity to meet demands that change throughout the day and year. Generally, the period of peak electricity use during the year is from November to March.

29. **Why would the use of electricity be so high in the November-March months?**

There is also a variation in demand for electricity on a daily basis. Power plants must be set up to meet the period of highest demand. When there is less demand, part of the power plant is not used fully.

Hydro-electric power is important in meeting peak period demand. When more electricity is needed suddenly, more water can be allowed through the turbines. This avoids the problem of starting up the boilers in a thermal electrical plant.

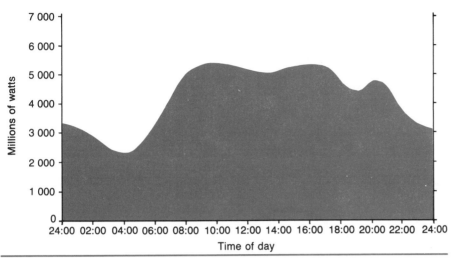

Figure 7-23 Variations in electrical demand for a typical summer day

30. (a) When is the least power demand?
    (b) Why?
31. (a) When is the greatest power demand?
    (b) Why?
32. (a) Look at the greatest and least power demands for this power plant (Figure 7.23). How many times greater is the peak than the least power demand?
    (b) In what specific ways could your household, your school, or nearby offices or factories help to change this uneven demand for electricity? Think of changing the times when electricity is used.

The North American Power Grid

Stretching across Canada is a vast network of high-tension power lines that carry our electricity. These lines transport electric power from the station to the markets. The network is so designed that if one power plant breaks down, electricity from other plants can supply the needed energy.

Long strips of land are reserved for these power lines. In urban centres a great deal of valuable land is used up for this purpose.

The large pattern of electrical lines in Canada is called a power grid. Our power grid is linked to that of the United States. This allows for transfer of electricity across the border from one country to another. Some provinces, such as British Columbia, Ontario, and Quebec, produce a surplus of electricity, which they can export to the United States.

33. What future problem might Canada face if agreements were signed to sell electricity to the United States for the next ten years?
34. What advantages would Canada gain from selling its electricity to the United States?

Figure 7-24 A power transmission line

Impact of Producing Electricity

Each method of generating electricity produces side effects that can be harmful to people and to the environment. When deciding on the most suitable method of production, it is also important to know how much of the energy is available to Canadians. The following table compares the various methods of electrical production.

SOURCE OF ENERGY	COAL	OIL	NATURAL GAS	URANIUM
Land pollution	1. Land dug up to mine coal 2. Solid waste from mining 3. Solid waste after burning of coal	1. Pipeline construction 2. Damage from construction of oil wells	1. Pipeline construction 2. Damage from drilling for gas	Radioactive pollution from mining waste
Water pollution	1. Hot water from power station put into lakes and rivers 2. Pollution from drainage of coal mines	1. Same as coal, No. 1 2. Chance of an oil spill in transportation	1. Same as coal, No. 1	Same as coal, No. 1
Air pollution	1. Ash, many gases 2. Acid rain	Many gases	Very little	Some radioactive gases
Effect on people	1. Breathing problems 2. Damage to crops from air pollution 3. Danger of serious accidents to lives of miners	1. Breathing problems 2. Damage to crops from air pollution	Almost none	Almost none except for the threat of an accident
Amount available for use	Large Canadian sources, e.g. Maritime Provinces and Alberta	Large foreign sources, limited Canadian sources at present	Large Canadian sources	Fairly large Canadian sources

Figure 7-25 A comparison of various methods of electrical production

35. Hydro-electricity is not included in Figure 7.25. From the discussion earlier in the chapter, list the problems associated with it. Use the factors in the table as a guide.

36. Which source of energy appears to be the most damaging to the environment? Explain your answer.

37. What specific advantages does hydro-electricity have over the other sources of energy?

38. In your opinion, which of these energy sources appears to be most suitable for Canadians? Give reasons to support your answer.

ENERGY ALTERNATIVES

Many people are becoming concerned that one day we will begin to run short of energy. To reduce the demand on non-renewable resources, other methods are being tested.

As we discussed earlier, it is best to rely on renewable energy sources because they will not run out. One of the most talked about renewable energy sources is solar power. Its advantages are obvious. The sun's energy is free and available year-round almost everywhere in the world.

Active Solar Heating

At present, solar energy is used in Canada for heating only a few homes. One system is illustrated in Figure 7.26. This method of heating is referred to as active solar heating.

Figure 7-27 A solar-powered house

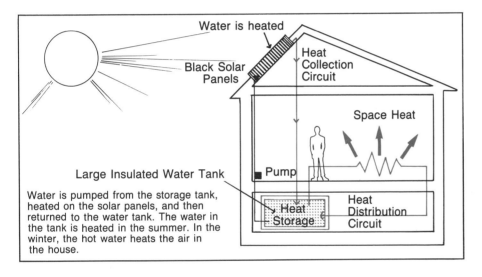

Water is heated

Black Solar Panels

Heat Collection Circuit

Space Heat

Large Insulated Water Tank

Pump

Water is pumped from the storage tank, heated on the solar panels, and then returned to the water tank. The water in the tank is heated in the summer. In the winter, the hot water heats the air in the house.

Heat Storage

Heat Distribution Circuit

Figure 7-26 Active solar heating

The first apartment building in Canada to be heated entirely by solar energy was opened in the summer of 1979. It is located near Aylmer, Ontario. Active solar heating is expensive to install, but is inexpensive to run. Experiments to improve the system are still continuing.

There are also devices called solenoids that convert solar energy to electricity. These devices are expensive, so that very few houses using solenoids exist at the present time.

One experimental house in Prince Edward Island, called The Ark, was designed to be heated with solar energy. It also has several windmills to provide some of its electricity. The Ark has attempted to be self-sufficient in energy but has been closed because of the costs of running it.

---

39. **Do you know of a building that uses active solar heating? Describe its location and explain what it is like.**

Passive Solar Heating

Passive solar heating is more widely used in Canada than the active type. Passive solar heating involves numerous small improvements in and around a house to save on heating costs.

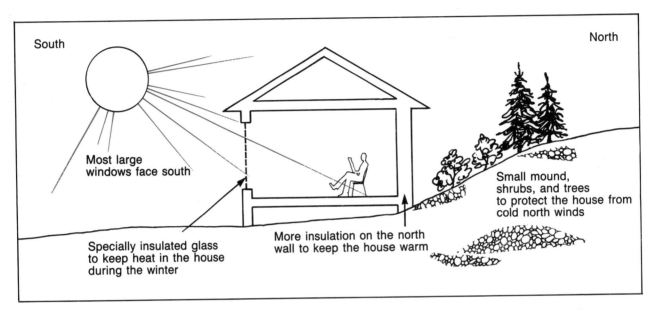

South

North

Most large
windows face south

Small mound,
shrubs, and trees
to protect the house from
cold north winds

Specially insulated glass
to keep heat in the house
during the winter

More insulation on the north
wall to keep the house warm

**Figure 7-28 Passive
solar heating for a
Canadian home**

The purpose of this activity is to plan an entire subdivision (development) of individual houses. Each house is to be designed to include passive solar heating features. Follow these steps to map out the location of your houses.

40. Obtain two blank sheets of paper, pencil, eraser, and ruler.
41. Draw a square in the centre of your first sheet of paper. Each side of the square should measure 15 cm. Place a north arrow to point to the top of your page.
42. On the square, mark eleven houses. Each house should be 1 cm², and have a street(s) to join them. This street must also run to the edge of your square to join with other, outside streets. Make the street 1 cm wide.
43. Once you have drawn in your houses and roads, begin to mark the passive solar-heating features, such as small mounds, trees, and shrubs. Remember the cold winter winds blow from the north.
44. Carefully examine all your houses to make sure each one has one wall facing south. Mark on the boundaries of the house lots.
45. Give your plan an appropriate name.
46. Now that you have finished your plan, design an advertisement for the houses you have planned. On your second sheet of paper set up an advertisement to be used in a local newspaper. Be sure to mention the passive solar-heating features the houses have.

Wind Power

Windmills take advantage of a free source of energy available all across Canada. Although wind is free, it does not blow all the time. Nor does wind blow with the same force all the time. As a result, wind power is suitable for only some regions of Canada, such as the Prairies or the Magdalen Islands. For the rest of Canada windmills could only be a supplement (help) to the sources of power we already have.

Figure 7-29 Using wind power

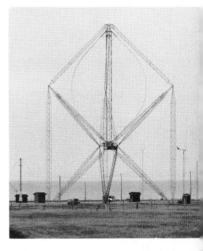

(a) Windmill

(b) New wind-powered generators

Plant Material as Fuel (Biomass Conversion)

Across Canada every spring, farmers plant hundreds of different kinds of crops. In the autumn these crops are harvested and the fields are cleared for the next planting in the following spring. If this kind of timetable can be used for crops we eat, why could we not use it for crops that provide energy? We should try to grow plants that could be harvested and then treated chemically to give us energy. This is called biomass conversion.

Trees are being used now to produce energy. Since trees are often considered a crop, they are cut when they are mature. After the wood is treated with chemicals, it can produce methanol, which is used as a fuel. The land that was cleared of trees can be replanted, so that perhaps forty years from now trees can be harvested again for energy. Trees that are not good enough for lumber could be used to produce energy.

47. What great advantages does biomass conversion have over windmills?
48. What undesirable side effects might biomass conversion have that both windmills and solar power do not have?

Tidal Power

In Chapter 6 you discovered that Canada had the longest shoreline in the world. Along part of the coastline there are **tides** large enough to produce power. Tides are the rising and falling of the water level on the ocean coasts.

The area that holds the greatest promise of tidal power is around the Bay of Fundy, where the greatest tides in the world are found. The water level can change up to 20 m between high and low tides.

Figure 7-30 The effect of tides in the Bay of Fundy *Note the pier and ship on the right.*

(a) High tide

(b) Low tide

As illustrated in Figure 7.30, tides contain a great amount of energy that is available every day. There are two high tides and two low tides per day. Figure 7.31 demonstrates how these tides could be used to generate electricity.

Figure 7-31 (a) Using tidal power to generate electricity

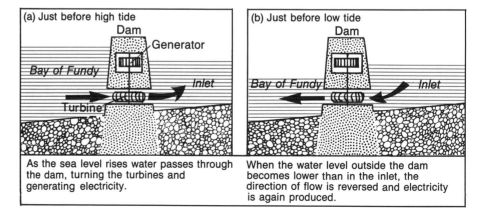

(a) Just before high tide

Dam

Generator

Bay of Fundy

Inlet

Turbine

As the sea level rises water passes through the dam, turning the turbines and generating electricity.

(b) Just before low tide

Dam

Bay of Fundy

Inlet

When the water level outside the dam becomes lower than in the inlet, the direction of flow is reversed and electricity is again produced.

Figure 7-31 (b) Tidal power station in Nova Scotia *The town of Annapolis Royal is just out of view.*

49. Look at Figure 7.31 (b) and describe how power is produced from the rising and falling of the tides.
50. Tidal power has already been used in several locations in France to produce electricity. Canada's first tidal power station is on the shore of the Bay of Fundy in Nova Scotia. Turn in your atlas to a map of the Bay of Fundy area. Examine the northern shoreline of the Bay. Locate those inlets from the Bay that are long and narrow but have a small mouth. These inlets would be well suited for tidal power.
    List the names or locations of these inlets.
51. In what specific ways would the production of tidal power differ from (a) hydro-electricity and (b) thermal electricity?

CONSERVATION

Most of our energy sources are non-renewable. More and more Canadians are using methods to reduce the amount of energy which they use. This will help each of us to reduce energy costs and ensure a future supply.

Energy conservation requires many changes in the lives of individual Canadians. Not only must Canadians think differently; they must also spend money to conserve energy. The key to energy conservation in Canada lies largely with you and your generation.

Transportation

In a country such as Canada, transportation is of vital importance. Canada is the world's second largest country and has a population which is very spread out. As a result, Canada relies heavily on its transportation network to knit it together as a nation.

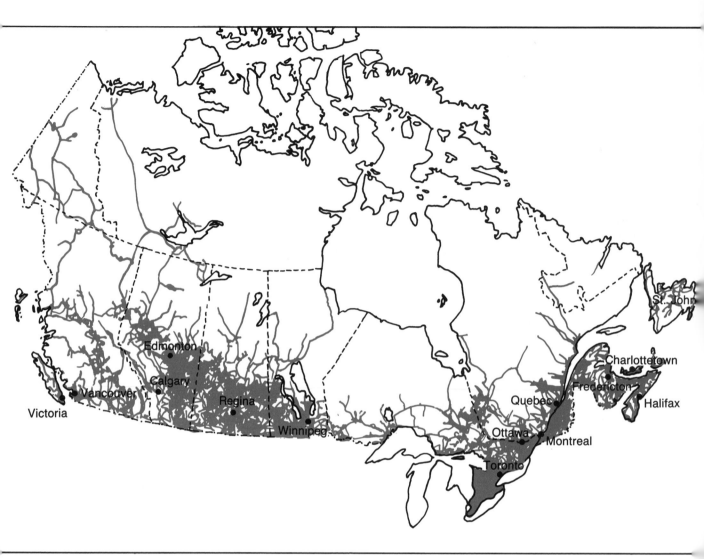

Figure 7-32 Road density in Canada

Using Figure 7.32 and a map in your atlas that shows Canada's major roads, answer these questions.

52. (a) What is meant by road density?
    (b) In which direction do most Canadian roads run?
    (c) How is population density related to the density of roads?

53. Select two of the following provinces:
    New Brunswick, Ontario, Saskatchewan, British Columbia.
    (a) Describe the road-density pattern in the provinces you have selected.
    (b) What is the relationship between the location of cities and the density of the road network?

54. Some transportation routes run into the Canadian North. Examine thematic maps of Canada in your atlas, and explain the purpose(s) of such routes.

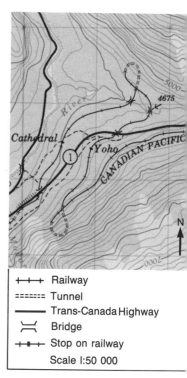

+—+—+	Railway
=======	Tunnel
——	Trans-Canada Highway
⨝	Bridge
+—■—+	Stop on railway
	Scale 1:50 000

Figure 7-33 Spiral tunnels near Banff, Alberta *These tunnels permit trains to negotiate this steep mountain pass. The train cars on the track are the tail end of the train coming out of the tunnel.*

Figure 7-34 A map of the spiral tunnels in Alberta

As you will discover in Chapter 8, Canada has large quantities of natural resources. Many of these resources are located far from the markets that need them. Therefore, although moving these resources is important, it is also costly. In fact Canada's physical features and climate make the building of transportation lines very difficult and expensive. Some of the greatest engineering feats (projects) in the world are located on Canadian railway lines and highways.

The spiral tunnel on the Canadian Pacific Railway Line in the Rocky Mountains near Banff, Alberta, is a good example of costly and sophisticated engineering.

In northern Ontario the Trans-Canada Highway was built across several very large swamps. This construction required extensive surveying, forest clearance, and rock fill before the road could be constructed.

55. Describe in detail a journey by train along the Canadian Pacific Railway line from east to west, through the area included on the map in Figure 7.34.
56. (a) What is the straight-line distance for the journey which you described in Question 55?
    (b) How far does the train travel?

DIFFERENT FORMS OF TRANSPORTATION

Methods of transportation used in Canada vary according to what is being carried. Coal, oil, people, and ice cream, for example, each requires a different method of transportation. To accommodate each of these commodities special consideration must be made.

---

57. Construct a table in your notebook similar to the one below to show how different commodities are moved.

COMMODITY	SPECIAL CONDITIONS FOR THE TRANSPORTATION OF COMMODITIES	TYPES OF TRANSPORTATION POSSIBLE (EXAMPLE: REFRIGERATED TRUCK)
Eggs People Ice cream Gasoline	**SAMPLE**	**ONLY**

You have discovered that transportation has become very specialized in Canada. As transportation methods improve more commodities can be moved. We now take for granted items such as frozen orange juice and fresh vegetables in the winter. Both require special transportation facilities.

TRANSPORTATION IN NORTHERN CANADA

One major problem with the development of the Canadian North is that of transportation. Minerals, oil, natural gas, and other resources could be moved from the Canadian Arctic to the market if adequate transportation was available. Most of the Canadian Arctic is so isolated that no highway or railway line comes close. Since the Arctic Ocean is frozen or blocked with ice most of the year, ship transportation is very difficult. One additional problem exists in the Arctic. Weather conditions are often unpredictable. Fog, storms, and strong winds can suddenly develop, making aircraft travel dangerous or impossible.

The obstacles mentioned make development of the Canadian Arctic a major problem. If somehow these problems could be overcome, the Canadian Arctic could be of great importance to the entire country.

At present there are two major highways built into the Canadian North or Arctic. These are the Alaska and the Dempster Highways. The Dempster Highway was officially opened in 1979. It is the first good road link to the delta of the Mackenzie River. Both roads are important to the resource development of the North. Logs, minerals, and other resources can reach the markets via one of these highways. Also, supplies can reach northern settlements more easily. This will allow the northern communities to grow in size.

TRANSPORTATION INSIDE A CITY

Most Canadian cities have a **grid pattern** for their streets. With such a grid pattern, the roads are straight and meet other roads at right angles. Figure 7.35 is a partial map of the street pattern of Edmonton.

Edmonton's street pattern was simple to set out. It provides direct routes across the city, without complicated intersections.

As you can see by the scale on Edmonton's map, the city sprawls across the landscape. The automobile, which is so popular among Canadians, allows cities to grow over a large area. Residents of Edmonton can travel by car to all parts of the city with relative convenience.

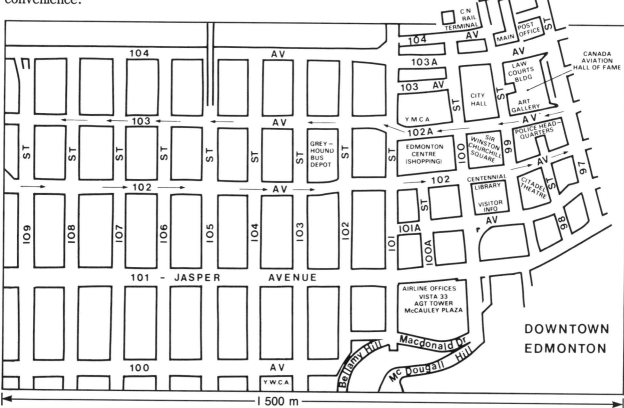

Figure 7-35 Part of a road map of Edmonton

58. Would Edmonton's grid-pattern streets be suitable for a tourist or new resident to find their way around? Explain your answer.
59. As was pointed out, Canadians are very attached to their cars. Give four good reasons to explain why this is true.

Public transportation via buses is also available along Edmonton's main streets. Although public transit is important, cars and private vehicles still provide most of the transportation. In fact 80% of all urban travel in Canada is by car. Approximately 60% of all transportation activity in our country takes place in urban areas such as Edmonton.

60. (a) What are the advantages of a good public transit service (buses and trains) in a city?
    (b) Why do so many people use cars in cities where there is a good public transportation system?
61. What suggestions can you make to decrease traffic-congestion problems in large cities?

TRANSPORTATION: KEY TO CANADA'S RESOURCES

Transportation and accessibility are very important for the development of Canada's resources. Although Canada has a rich variety of natural resources, they are often located too far away from the markets for full development. Examine Figure 7.36. Take special note of the locations of the resources.

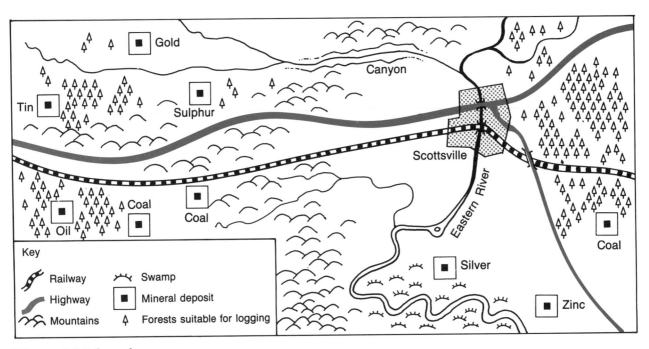

Figure 7-36 Area for resource development

62. (a) Which three resources should be developed first? Give reasons for your answer.
    (b) What transportation routes might be built to develop these resources? What specific problems would there be in building these transportation routes?
    (c) Which three resources do you feel are not worth developing at present? Explain the reasons for your answer.
63. Write one-third of a page to explain the importance of transportation for the development of Canada's natural resources. Refer in part to your answer to Question 62.

Petroleum
Distillation
Cracking
Seismograph
Continental shelves
Oil sands

Boom town
Megajoules
Hydro-electricity
Thermal electricity
Nuclear reaction
CANDU reactor

Radioactivity
Power grid
Biomass conversion
Tides
Grid pattern

Answers to the Energy Quiz on page 208.
1. (a) 2; (b) 2; (c) 5; (d) 2
2. (a) 3; (b) 5; (c) 0
3. (a) 5; (b) 0
4. (a) 3; (b) 1
5. (a) 3; (b) 0; (c) 3
6. (a) 0; (b) 3
7. (a) 0; (b) 3
8. (a) –3; (b) –4 (c) –2; (d) –5 (Subtract)
9. (a) 3; (b) 5 (c) 0
10. True 0 False 5

If you scored 25 or over, you are concerned about our energy supplies. If you scored 15-25, you are an average citizen and should start to become more aware of the problems. If you scored lower than 15, you should start to change your ideas now.

1. In what specific ways could your school save energy and yet still operate a full program? Examine such areas as windows, doors, lights, heating, and use of paper, etc.
2. Gather together copies of your local newspapers as well as popular magazines and advertising pamphlets that have been delivered to your home. Search through this information for devices that are designed to save energy. How is each device supposed to save energy?

8 OUR NATURAL RESOURCES AND HOW WE USE THEM

Our Natural Resources

Canada is known throughout the world as a source of industrial raw materials. These raw materials come directly or indirectly from our **natural resources** of rocks, trees, water, soil, and wild animals. A natural object becomes a resource only when we begin to use it.

Natural resources fall into two main groups — renewable and non-renewable. **Renewable resources** are those that can replace themselves once they are used. Forests, when managed properly, for example, can regrow after cutting. In contrast, once **non-renewable resources** are used, they are said to be **exhausted,** since they cannot replace themselves. Iron ore, oil, and coal are examples of non-renewable resources.

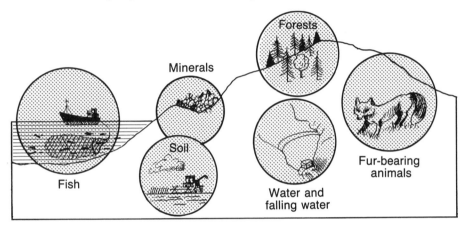

Figure 8-1 Natural resources *Natural things become resources when we learn to use them profitably.*

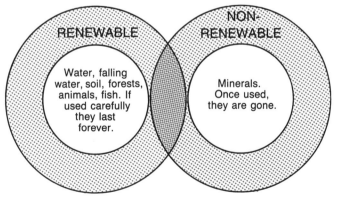

Figure 8-2 Renewable and non-renewable resources *Notice that the two groups overlap. Living things can be killed off if we misuse them. Minerals are always being formed, but not as quickly as we use them.*

1. (a) What natural resources are being *exploited* (used) in an area within 10 km of your school?
   (b) List any resources that have been exhausted in the same area.
2. (a) Explain why water is considered to be a renewable source.
   (b) Suggest factors that might cause water to be exhausted.
3. In what specific ways might the following be considered natural resources?
   • wildflowers   • mountains   • tree stumps

Our natural resources have been of key importance in attracting settlers to Canada. The rich fishing grounds off the coast of Newfoundland drew many Europeans to Canada's east coast, starting in the early 1500s. The interior of Canada was first mapped by European fur traders as they travelled across the country. Even the gold rushes of western Canada drew many to our country.

The search for new wealth continues today. When resources are discovered, their exploitation may require hundreds of workers, transportation routes for resources to reach the markets, and supplies to reach the workers. Whole towns are sometimes built in a previously uninhabited area, as we discussed in Chapter 7.

Thompson, Manitoba, was built to house workers and their families after a large nickel deposit was found in the area.

Figure 8-3 The early development of our natural resources *Workers at a logging camp.*

Figure 8-4 Thompson, Manitoba *The mine and smelter appear in the background.*

4. Look at the photograph of Thompson (Figure 8.4). How can you tell that this is a planned town?
5. Why do you think that there are no tall buildings in Thompson?
6. Thompson was built 640 km north of Winnipeg, in an area of forest and lakes. It was 50 km from the nearest rail line.
   (a) What must have been done before the town, mine, and smelter were started?
   (b) What would a mining company have to do to attract workers to a remote centre such as Thompson?

Mining

The mining industry is important to each of us, as well as to Canada as a whole. Almost every industry depends on minerals to a greater or lesser degree.

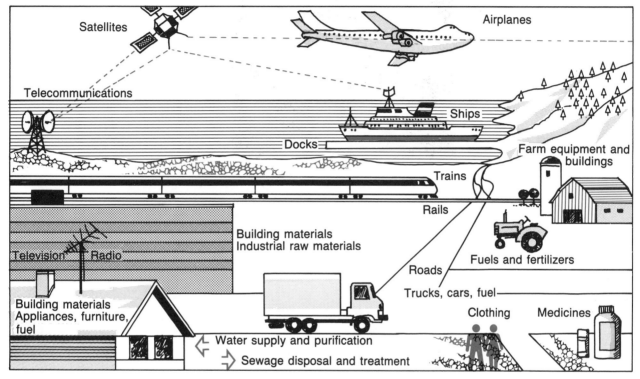

Figure 8-5 How our lives depend on minerals *This diagram shows a few of the ways in which we use minerals. Can you name some others?*

7. Using Figure 8.5 as a guide, write at least one page explaining how the way you live depends on products that come from minerals. You could do this quite easily by going through a typical 24 h period and noting your activities. Include references to such items as your food, clothing, and shelter.

THE IMPORTANCE OF MINING TO CANADA

Canada produces about 50 different mineral commodities. In 1983 these were valued at $36 000 000 000 and they made up 12% of the value of our Gross National Product. We exported $17 350 645 000 worth of minerals in 1984, which comprised 20% of the total value of our exports.

Mining creates much employment, both directly and indirectly. In 1982 approximately 120 000 were directly employed. This was 2% of the total labour force and included 5 000 students in summer jobs. Figure 8.6 will help you to list some specific types of employment resulting from mining activities.

Mining also supports many other industries. For example, moving crude and processed minerals makes up...

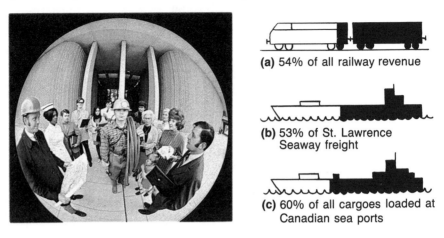

(a) 54% of all railway revenue

(b) 53% of St. Lawrence Seaway freight

(c) 60% of all cargoes loaded at Canadian sea ports

Figure 8-6 Mining creates many jobs

THE DISTRIBUTION OF MINING IN CANADA

Mining is of importance to all parts of Canada. An atlas map of natural resources will illustrate this fact. Figure 8.7 summarizes the importance of each province or territory in the production of minerals.

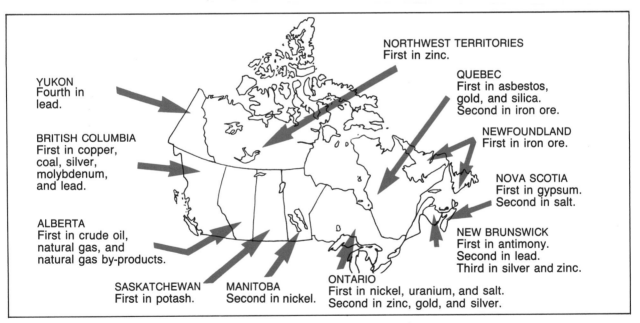

NORTHWEST TERRITORIES
First in zinc.

QUEBEC
First in asbestos, gold, and silica.
Second in iron ore.

NEWFOUNDLAND
First in iron ore.

NOVA SCOTIA
First in gypsum.
Second in salt.

NEW BRUNSWICK
First in antimony.
Second in lead.
Third in silver and zinc.

YUKON
Fourth in lead.

BRITISH COLUMBIA
First in copper, coal, silver, molybdenum, and lead.

ALBERTA
First in crude oil, natural gas, and natural gas by-products.

SASKATCHEWAN
First in potash.

MANITOBA
Second in nickel.

ONTARIO
First in nickel, uranium, and salt.
Second in zinc, gold, and silver.

Figure 8-7 The major contribution of each province or territory to Canada's mineral production

The mineral deposits found in an area depend upon the original nature of the rocks and what has happened to them since they were formed. In the folded mountain areas of the west and east, the minerals consist mainly of coal and metals such as lead, silver, zinc, and copper. In the fairly flat areas of the Interior Plains and the Great Lakes-St. Lawrence Lowlands, oil, gas, and non-metallic minerals such as gypsum, potash, and salt are being mined. The Canadian Shield contains mainly metallic deposits.

LOCATING VALUABLE DEPOSITS

In the early days of mineral exploration prospectors would travel through areas of dense bush, mosquito-infested swamps, high mountain ranges, and the icy North in search of valuable deposits.

Their equipment was primitive, a hammer or a gold pan being their most important items. They listened to tales told by the Indian people. Sometimes they were successful in their search, but often they were not.

Some major discoveries occurred by accident. In 1883, the rock near Sudbury, Ontario, was being blasted for the Canadian Pacific Railway. The blasting exposed part of the great nickel and copper deposit which today forms a vital part of the Canadian mining industry.

Figure 8-8 Prospecting many years ago

Figure 8-9 Modern methods used to determine the presence of ores (a) (b) (c) (d) (e) *The techniques used to find ores include measurements made from the air and oceans, and the sampling of soil and vegetation. Sophisticated equipment aboard ships and aircraft and in laboratories help in the analysis of the findings.*

243

Figure 8-10 A geologist examining a core sample

Prospecting methods have changed considerably since those early days. Frequently, valuable deposits are buried deep in the ground, sometimes even beneath lakes and oceans. Such ore bodies are located by scientists using special instruments on the ground, suspended from aircraft or attached to ships. These methods aid experts in determining the shape of the rocks beneath the surface, and the presence of ore bodies. An **ore** is a mineral deposit that is worth mining.

When the scientists believe that they may have found an ore body, they drill into the rocks. From this drilling, core samples are obtained. The **core samples** are long cylinders of the rocks beneath the surface. These samples are used to determine whether the deposit is large and concentrated enough to mine.

DEVELOPING A MINE

If a mining company decides to proceed with mining, it has to raise capital (money) and buy land and **mineral rights** (the legal right to mine). It must also obtain a licence to mine and provide transportation from the mine to the marketplace. Workers must be hired and be provided with a place to live. Hundreds of millions of dollars may be invested in developing a mine, and even a large company cannot afford to make a mistake.

There are two major methods of mining ore. Where the deposit is close to the surface, **strip** or **open pit mining** is employed. When the deposit is deep in the ground, **underground** or **shaft mining** is used.

In open pit mining, the useless **overburden,** or covering layer of soil and rock, is scraped or blasted away and the ore is exposed.

The open pit is then created by blasting the uncovered ore. The inside of the pit is gradually carved into a series of steps, as you can see in Figure 8.12.

Figure 8-11 Scraping off overburden for open pit mining

Figure 8-12 An open pit mine

8. Examine Figure 8.12. What type of vegetation was probably growing at the mine site before it was developed?
9. (a) Why is an open pit mine arranged in steps?
   (b) Which levels are currently being mined? How can you tell?
   (c) Which level was mined first? Why was this?
10. (a) What environmental problems would exist after a company stopped using an open pit mine?
    (b) What could be done to make a disused mine more attractive?

Underground mining is more costly and dangerous than open pit mining.

Figure 8-13 The Falconbridge nickel mine and mill in Sudbury, Ontario

11. What are three major differences you can observe between the underground mine in Figure 8.13 and the surface mine in Figure 8.12?

The structure of a typical underground mine is illustrated in Figure 8.14. A vertical **shaft** is sunk from the shaft house. Miners and ore are raised and lowered along this shaft. Horizontal or near-horizontal **drifts** are **excavated** (dug) through the ore. These drifts often open up into larger **stopes.**

12. (a) Using Figure 8.14, write a detailed account describing how loosened ore gets from stope 8A to the surface.
    (b) How does the way in which the ore travels to the surface from stope 8B differ from the route you described in answering 12 (a)?
13. Explain at least two reasons why underground mining is more dangerous than surface mining.

ILLUSTRATION OF
AN UNDERGROUND MINE
1 HOISTING CABLE
2 CAGE
3 SKIP
4 SUMP
5 CRUSHER
6 LOADING POCKET
7 ORE PASS
8 STOPE
9 CROSSCUT
10 VENTILATION SHAFT AND
   SECONDARY TRAVELWAY
11 VENTILATION FANS

Figure 8-14 The structure of a typical underground mine

In the mill, which is usually situated close to the mine, the ore may be sorted, crushed, concentrated, and washed before being sent out to the manufacturers. Concentration of the ore cuts transportation costs because the worthless portion of the ore has been discarded. The concentrated ore may be crushed and formed into **pellets.**

Figure 8-15 Pelletized iron ore

Figure 8-16 Drilling the wall of a stope in preparation for blasting

Figure 8-17 Chemical fertilizer *Chemical fertilizer is an important product of the mining industry.*

Case Study: Potash in Saskatchewan

Potash is a key ingredient of most modern fertilizers used on farms and on gardens and lawns around Canadian homes. In fact, most crops grown in Canada would suffer greatly without the fertilizers in which potash is found. About 96 per of all the potash in the world is used in fertilizers. The rest is used in chemical and associated industries.

Fortunately, Canada is a major world producer of potash. Of all the provinces, Saskatchewan dominates in potash production. In 1984 Canada sold potash worth $759 270 000.

The first potash was discovered in Saskatchewan in 1942, but major production did not occur until 1962. A worldwide recession in the early eighties caused demand for potash to decrease. Only 12% of this is used in Canada and the rest is exported to other countries. Figure 8.18 compares Canada to other major potash producers of the world.

246

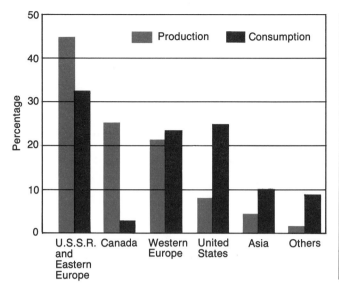

Figure 8-18 World potash production and consumption (1980)

Figure 8-19 Location of potash reserves and mines in Saskatchewan and surrounding areas

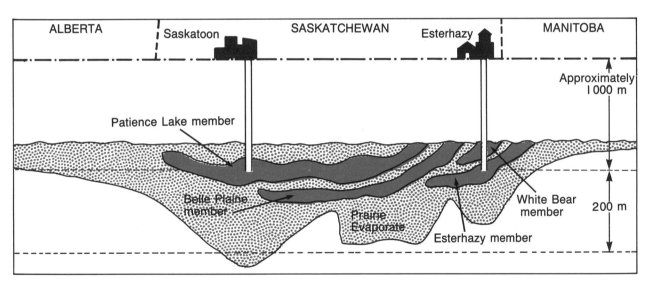

Figure 8-20 Cross section of potash beds in Saskatchewan

14. Examine Figure 8.18 closely.
    (a) List in order (highest to lowest) the five major areas of potash production in the world.
    (b) Which areas or countries have the greatest amount of potash available for export? Explain how you arrived at your answer.
15. (a) Which area is the greatest net importer? (A *net importer* imports more than it exports.) From which country would it import most of this potash?
    (b) Give two reasons why you think the U.S.A. uses more potash than Canada. (Refer to an agricultural map of North America in your atlas.)

16. In order to use potash fertilizer successfully, people in a country must have enough money to buy it and knowledge of how to use it.
    (a) Using information on income per person (per capita Gross National Product) from your atlas, list ten countries that would probably not use much potash.
    (b) Are your findings in 16 (a) in agreement with the facts illustrated in Figure 8.18? Explain how you obtained your answers.

The area in Saskatchewan with potash deposits is shown in Figure 8.19. The potash was formed beneath an inland sea about 360 000 000 years ago. An **inland sea** is one into which rivers flow, but from which little water escapes. The water in this sea slowly evaporated, leaving alternating layers of salt and potassium on its floor. Figure 8.20 shows the layers of potash.

Sometimes potash is obtained from underground mines similar to those described earlier.

Figure 8-21 Machine cutting potash deposits

When potash deposits are located more than 1 100 m below the surface of the ground, solution mining is used. (See Figure 8.22.)

17. Considering the nature of potash, why would underground mining be restricted to depths of less than 1 100 metres?
18. Why must potash be transported in waterproof containers?

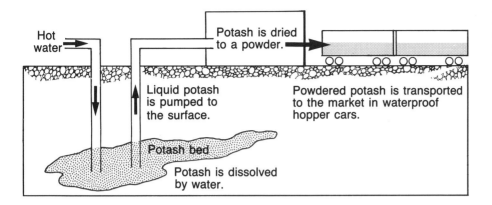

Figure 8-22 Flowchart of solution mining of potash

Hot water →

Potash is dried to a powder. →

Liquid potash is pumped to the surface.

Powdered potash is transported to the market in waterproof hopper cars.

Potash bed

Potash is dissolved by water.

Potash mining has resulted in many benefits, besides the money which comes into Canada each year. For example, 4 000 people are employed directly in the industry, while 10 000 others owe their jobs indirectly to the industry.

In the early 1960s too much potash was produced in the world. As a result, the mining companies cut their prices in an attempt to sell more. Unfortunately, the price of the potash fell below the actual costs of production. The mines had to then cut back on their output. This resulted in the loss of many jobs.

The provincial government at that time decided to limit the amount of potash produced by each mine. This policy is called **prorationing.** The government also set a **floor price,** which was the lowest selling price that would be accepted for potash. These two policies have now become common for other commodities. There are still conflicts over issues such as **royalties,** which are taxes that the mines must pay to the government. Royalties are based on the amount of potash produced at each mine.

The future for the potash industry of Saskatchewan looks good. One reason for this is the anticipated demand for potash from Third World nations where new agricultural methods are being introduced. However, other countries will be competing with Saskatchewan for these new markets.

THE IMPORTANCE OF CANADIAN MINING PRODUCTS TO THE U.S.A.

The U.S.A. is Canada's largest market, buying 62% of our exports of minerals and their **fabricated** (partially processed) products in 1984. Including oil and gas exports, all of which go to the U.S.A., the total value of these imports was $10 567 570.

19. Many people believe that Canada should not sell its natural resources to the U.S.A. and other countries. What is your opinion? What are the advantages and disadvantages of selling our minerals abroad?

SOME PROBLEMS ASSOCIATED WITH MINING

Figure 8-23 Some major mining problems

Any exploitation of natural resources usually results in pollution. Mining is no exception.

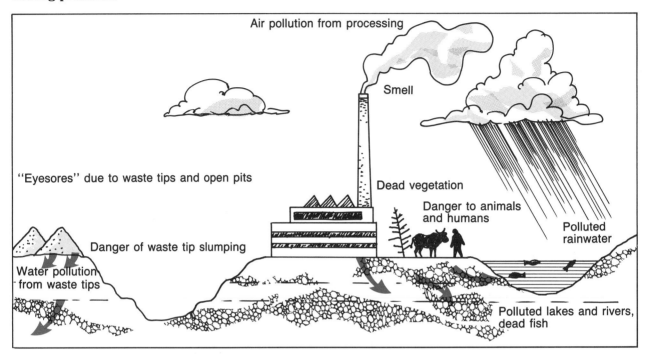

Air pollution from processing

Smell

"Eyesores" due to waste tips and open pits

Dead vegetation

Danger to animals and humans

Polluted rainwater

Danger of waste tip slumping

Water pollution from waste tips

Polluted lakes and rivers, dead fish

Figure 8-23 Some major mining problems

Figure 8-24 Reducing air pollution *Trail, B.C. on the Columbia River is the site of the world's largest lead-zinc smelter. Photograph (a) was taken in 1930. There was no vegetation due to sulphur in the smoke. In that year Cominco began building plants to remove the sulphur for use in the manufacture of chemical fertilizers, at that time virtually unknown in Canadian agriculture. As the air was cleaned, Cominco planted 2 000 000 trees over a period of years. Today there is less sulphur in the air at Trail than in many large cities. Vegetation flourishes, and Cominco meets the rigid environmental control rules imposed by government — while producing more than 1 000 tonnes per day of lead and zinc.*

Pollution can be controlled, but the processes are expensive. Because pollution control is so expensive, laws have been made to force companies to reduce the amount of pollution they create. The extra expense of pollution control makes our exports cost more. Sometimes we cannot sell our products abroad because they cost more than those from another country.

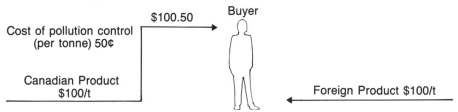

This is true for many other Canadian industries. Pollution control is expensive. We will have to choose between the environmental costs of pollution and the costs of fighting pollution.

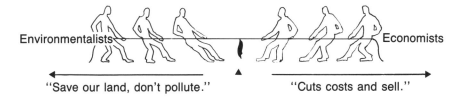

Figure 8-25 Reducing pollutants in water

20. If you live in a mining area, or have visited one,
    (a) describe problems associated with mining and the environment in that area
    (b) describe any steps taken to reduce the damage.
21. You will soon be old enough to vote for members of local councils and government representatives. Write a letter to one of these people. In your letter, write down your opinions about how mineral development should be controlled to protect the environment.

Forestry

Canada is a land of forests. About 3 000 000 km² of trees, covering over one-half of Canada's area, extends in a great curve from the Atlantic to the Pacific. Trees provide us with many products, some of which are shown in Figure 8.26.

Only 2 300 000 km² of forests, mainly in the south, is suitable for profitable use. The factors that reduce the value of a forest are as follows:
• a cold climate, which only allows low tree growth
• too dry a climate
• poor drainage as in marshes, or rocky, poor soils
• lack of **access** (the ability to reach) to an area

Figure 8-26 Some of the many products derived from a tree

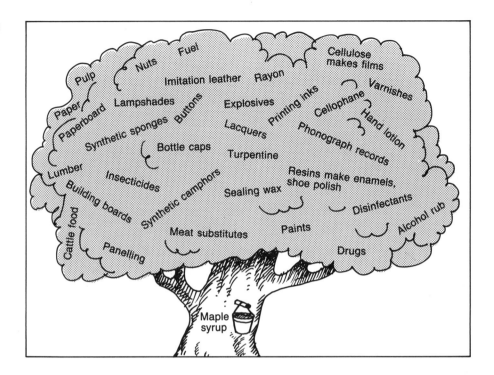

22. Refer to the forestry map in your atlas. Which parts of Canada have little or no forest resources suitable for exploitation?
23. Using the answer to Question 22 and climate and transportation maps in your atlas, explain why these areas have such poor forest resources.

THE IMPORTANCE OF CANADIAN FORESTS

Canada contains about 10% of the productive forests in the world. Pulp and paper is Canada's largest manufacturing industry. Each year Canada exports over $15 000 000 000 worth of newsprint, wood pulp, and other paper products to over 90 countries. This makes up 13.9% of our total export earnings. Nearly 150 000 people are employed in the forestry and forestry manufacturing industries. Indirectly, hundreds of thousands of additional people are employed. For example, the pulp and paper mills use one-eighth of all of the electric power consumed in Canada. Also, every eighth freight car on our railways contains pulp or paper products.

FOREST INVENTORY

To determine as accurately as possible the quantities of available timber, an **inventory** (counting) was undertaken several years ago. Forest inventories include the techniques shown in Figure 8.27.

The results of the forest inventory (Figure 8.28) show clearly that British Columbia has our greatest timber reserves. Ontario and Quebec also have large reserves.

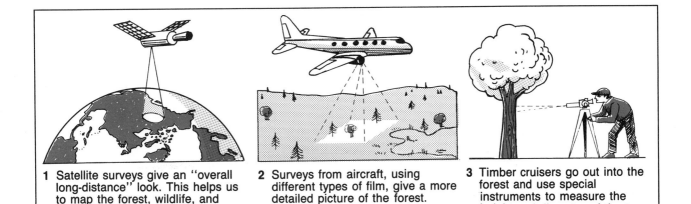

1 Satellite surveys give an "overall long-distance" look. This helps us to map the forest, wildlife, and water resources.

2 Surveys from aircraft, using different types of film, give a more detailed picture of the forest.

3 Timber cruisers go out into the forest and use special instruments to measure the height and diameter of the trees.

Figure 8-27 Some methods used in a forest inventory

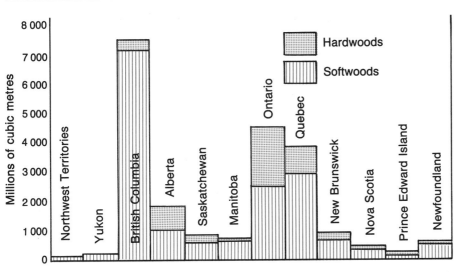

Figure 8-28 Volume of wood by province or territory

The trees are usually divided on the basis of their use.

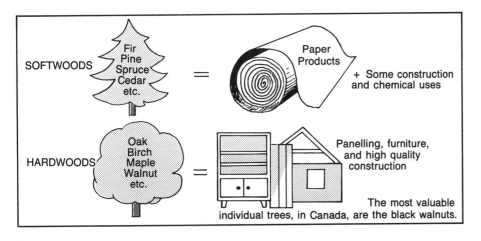

SOFTWOODS
Fir
Pine
Spruce
Cedar
etc.
= Paper Products + Some construction and chemical uses

HARDWOODS
Oak
Birch
Maple
Walnut
etc.
= Panelling, furniture, and high quality construction

The most valuable individual trees, in Canada, are the black walnuts.

24. **Refer to Figure 8.28 and write three-quarters of a page on the distribution and type of timber resources in Canada.**

MODERN METHODS OF FOREST MANAGEMENT

As Canadians, we are beginning to realize the many ways in which we benefit from our forests. These are illustrated in Figure 8.29.

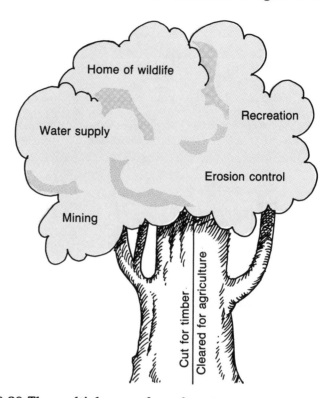

Figure 8-29 The multiple use of our forests

---

25. **List, in order of their importance to you, the various uses of our forests. Give reasons to explain your ranking of these uses.**

We are also learning that we must manage our forests as carefully as a farmer manages the land. Under proper management, areas that have been cut must be replanted or reseeded. The trees must also be protected from insects and disease.

Forest fires should be prevented when possible, and fought when they endanger life, property, communications, or valuable timber reserves. By controlling these threats and replanting to ensure future timber supplies, the forests are considered a crop to be harvested.

Harvesting the forest may be done in one of two basic ways — clear cutting or selective cutting. **Clear cutting** involves the removal of all trees suitable for logging in an area. In a well-managed logging operation, the old stumps, brushwood, and weeds are removed by **prescribed burning**. Fires are set out by forestry workers when conditions are suitable. The seeds or tree seedlings are then planted.

Figure 8-30 The spruce budworm has destroyed vast areas of our valuable spruce forests

Figure 8-31 Strip clear cutting

Figure 8-32 Selective cutting

Figure 8-33 Replanting the forest *The seedling trees ready for planting are shown on the left.*

26. Not all trees that are cut are replaced by planting. What steps could the government take to ensure that trees regrow once a forest area has been cut down?
27. Compare forests to regular field crops such as wheat. How are forests different from and how are they similar to these crops?

Fairly flat areas that are to be clear cut are often done using huge machines such as that shown in Figure 8.34.

When only certain trees are to be cut, as in selective cutting, the logger uses only a chain saw. Chain-saw cutting is also important where the land is too steep for large machinery.

28. For what purposes would selective cutting be practised?
29. What advantages and disadvantages are there in using large machinery for cutting the forests?

Figure 8-34 A large machine used in clear cutting the forest

Figure 8-35 Collecting logs together in British Columbia

To be able to get the machinery and people in and the logs out of timber-producing areas, logging roads, camps, and sometimes even towns like Kapuskasing and Fort Frances in Ontario must be built. In fairly flat areas, **skidders** (special kinds of tractors) are used to drag the cut logs to the road. In steep areas, especially British Columbia and Newfoundland, logs are dragged to a central spar (a tall tree or metal pole), using cables and a pulley system.

Experiments have been carried out with helicopters and helium-filled balloons to move logs. Helicopters are good, but are expensive to use. A practical and successful method using balloons has not yet been worked out.

TRANSPORTING THE TIMBER TO THE MILL

The methods used to move timber downstream vary from place to place. They are illustrated in Figures 8.36, 8.37, and 8.38.

In British Columbia most logs are transported to water or to the mill by huge trucks. Some of these trucks will carry up to 180 tonnes each. East of the Western Mountains 44% of the wood travels at least part of its journey by water. The largest river drive in the world is on Quebec's St. Maurice River.

More trucks are now used for moving logs. The number of logs going by water and rail is decreasing.

256

Timber is cut

Cut into lengths

Moved to a loading area by skidder

Loaded onto trucks

Unloaded into river and floated downstream

Made into rafts

Taken to mill

The methods used to transport timber vary from place to place.

The loading area for the logs may be by a road, a river, a lake, or by a railway. Floating logs down river is still very important.

Figure 8-36 Flowchart showing how logs reach the mill

Figure 8-37 Towing log rafts in sheltered waters

Figure 8-38 A self-loading, self-dumping barge in British Columbia

Figure 8-39 The St. Maurice River drive *The largest river drive in the world is in Quebec's St. Maurice River. More than three and a half million cubic metres of wood move downstream each year to mills at La Tuque, Grand'Mere, Shawinigan, and Trois Rivières. Lumber and pulp mills are always located between the forest and the market for forest products.*

FOREST FIRES

Every summer fires destroy tens of millions of dollars worth of timber. They threaten life, property, and transportation links. The two main causes are human carelessness with matches, cigarette butts, and campfires, and lightning, which often strikes in inaccessible areas. Firefighting and prevention are an important part of modern forest management.

A number of firefighting techniques are illustrated in Figure 8.40.

(a) Building firebreaks

(b) Using hoses

(c) Air-dropping water

(d) Dropping chemicals to slow down the movement of flames

Figure 8-40 Methods used to fight forest fires

The key to effective fire-fighting is early detection of fires. When fires are small it is easier for water bombers and fire crews to suppress them.

The following precautionary measures are taken:

- A lightning locator network pinpoints lightning strikes. These are displayed on a television monitor at the fire control headquarters. Aircraft are sent to places where lightning has struck, to see if a fire has resulted.
- In western Canada fire towers are still used to watch for smoke.
- Computers are used to keep track of temperature, humidity, winds, rainfall, the types of trees, and the condition of the forest. This information helps fire managers to work out the best way to fight a fire.
- Computers are also used to determine the fastest way to get people and equipment to the fire site. Fire personnel are sent to the fire scene by road, plane, or helicopter.

- Burning permits may be withdrawn, and forests may be closed to travellers and workers if the forest fire danger is too high.
- Forest resource companies train their workers to act as firefighters in times of fire emergency.
- Approximately 70% of all forest fires are the result of human carelessness. This means that 70% of all forest fires can be prevented if we use caution and common sense.

30. List the natural conditions and human activities that could lead to forest fires.
31. In 1980 in Ontario alone, $57 000 000 was spent to fight forest fires. Damage to timber was over $1 000 000 000. Such fires also damage animal and fish habitat, towns, villages, and isolated homes, and spoil the environment for tourists. Why is it considered worthwhile to fight forest fires?
32. Explain any ideas you may have for reducing forest fire losses in Canada. You may wish to do this in groups and present your conclusions to the class.

Figure 8-41 A fire watch tower *Fire towers such as this one are still in use in western Canada.*

THE PULP AND PAPER INDUSTRY

Pulp and paper is Canada's leading manufacturing industry. Its products make up 8% of our exports and are valued at over $7 billion annually. We produce 40% of the world's supply of newsprint. In addition, each Canadian uses an average of 140 kg of paper each year.

Figure 8-42 Some of the many paper products

There are about 150 mills in operation. The requirements for the production of pulp and paper include timber, fresh water, chemicals, and workers.

Figure 8-43 The requirements for a pulp and paper mill

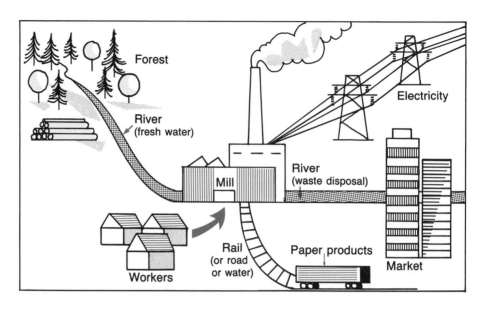

Once the logs arrive at the mill, they go through the series of processes illustrated by the flow chart in Figure 8.44. Paper making depends upon the fact that wet wood fibres will stick to one another as the water is removed.

Figure 8-44 The stages in the production of paper

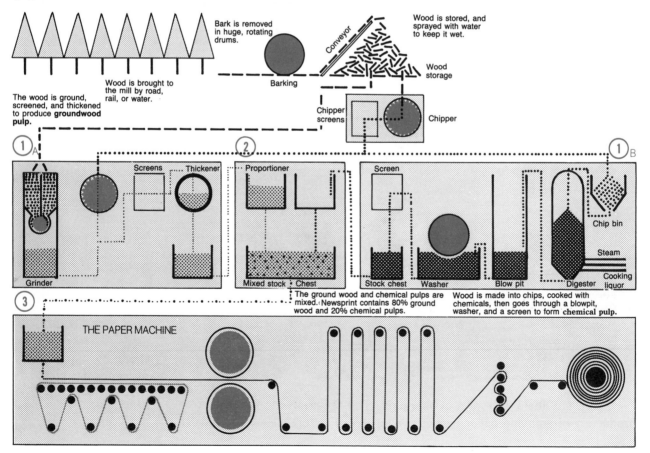

Figure 8-45 The paper machine

The map (Figure 8.46) shows the distribution of pulp and paper mills in Canada. While most of the mills are located close to the source of timber, some are found close to their markets, which is another important consideration.

Quebec is the largest producer of pulp and paper, accounting for 33% of the total production. British Columbia and Ontario produce 24% and 21% respectively.

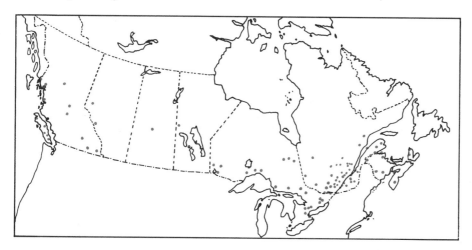

Figure 8-46 The distribution of pulp and paper mills in Canada

33. Make a list of 10 different types of paper products that you use. Describe the characteristics of the paper in each product. These characteristics could refer to the feel, colour, strength, and purpose of the paper.
34. Use Figure 8.44 to answer these questions.
    (a) List the stages at which water is used in the production of pulp and paper.
    (b) Explain why large quantities of electricity are required to make paper.

35. Select two other people from your class to work with you. The three of you have been appointed as consultants to the government for the remote area shown in Figure 8.47.

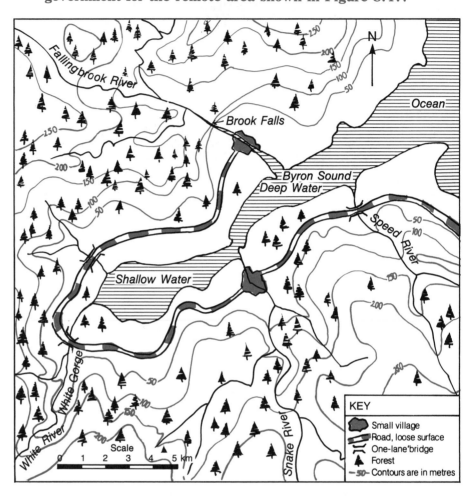

Figure 8-47 Area for pulp and paper developments

Your job is to attract three pulp and paper mills to the area.
(a) Trace the map onto a piece of paper.
(b) Carefully select sites for three pulp and paper mills. Keep in mind the requirements for a pulp and paper mill shown in Figure 8.43.
(c) Mark onto your map other facilities that would need to be developed. (Examples: hydro-electric dams, reservoirs for water, hydro lines, towns, roads, rail lines, etc.) Include a key to explain the symbols that you use.
(d) Give your map a suitable title and mark on a north arrow.
(e) Write an account to accompany your map. It should be in the form of a report to the directors of the pulp and paper industry and should carefully explain
(i) the advantages of the sites you have suggested for each mill
(ii) the reasons for the other developments.

Environmental Problems Associated with the Forestry Industry

As was the case with the mining industry, forest companies add a significant amount of pollution to our environment.

There are some solutions to reduce or eliminate these problems, as you can see in Figure 8.48. Research is constantly being carried out to reduce environmental hazards, but the solutions are expensive and take time to develop and install. Over a billion dollars has been spent by the pulp and paper industry to reduce air and water pollution. There have been significant improvements, but more still remains to be done.

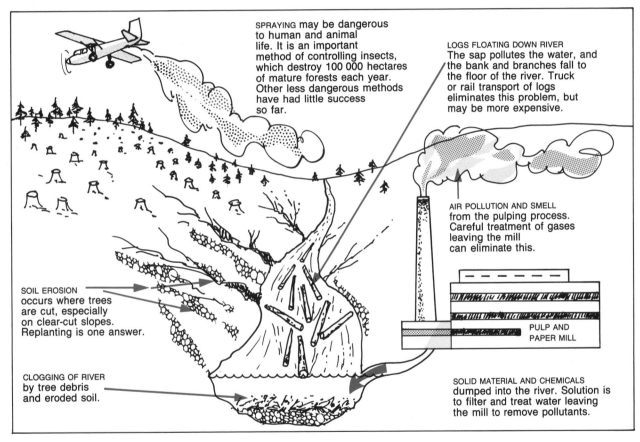

SPRAYING may be dangerous to human and animal life. It is an important method of controlling insects, which destroy 100 000 hectares of mature forests each year. Other less dangerous methods have had little success so far.

LOGS FLOATING DOWN RIVER The sap pollutes the water, and the bank and branches fall to the floor of the river. Truck or rail transport of logs eliminates this problem, but may be more expensive.

AIR POLLUTION AND SMELL from the pulping process. Careful treatment of gases leaving the mill can eliminate this.

SOIL EROSION occurs where trees are cut, especially on clear-cut slopes. Replanting is one answer.

CLOGGING OF RIVER by tree debris and eroded soil.

PULP AND PAPER MILL

SOLID MATERIAL AND CHEMICALS dumped into the river. Solution is to filter and treat water leaving the mill to remove pollutants.

Figure 8-48 Problems associated with the forestry industry

One very significant event in the reduction of pollution has been achieved in a pulp and paper mill at Thunder Bay, Ontario. The mill started operation in 1976, and is gradually coming closer to being a "closed-circuit" system. When it becomes a closed-circuit system, there will be virtually no pollution allowed into the environment. Chemicals and water are reused, and solid wastes such as bark are burned to generate electricity. This mill has not yet become a perfect closed-circuit system, but pollution levels from it are very low. Experts from all over the world study its operation to see if they can use the same ideas in their own countries.

Figure 8-49 The new "closed system" mill at Thunder Bay, Ontario
This is the first closed-cycle bleached kraft pulp mill in the world. It is designed to recycle liquid process wastes within the mill and discharge essentially clean water, used for cooling purposes, to the waterway.

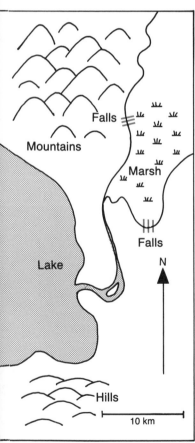

Figure 8-50 The area to be developed

36. Choose a partner to work with. Imagine yourselves as government planners, whose job is to decide how to make best use of the area of land illustrated in Figure 8.50.

    The following groups of people have presented arguments explaining why the area should be used according to their wishes.

    Forestry industry: It would develop the economy of the area and of the country, and would provide employment for many people.

    Preservationists: The area is a beautiful, undeveloped place that should be preserved in its present state. No hunting should be permitted.

    Tourists: Tourists would like to see towns, and recreational and skiing facilities developed. This would also develop the area economically.

    Railway: The railway company would like to be able to build a track through the area, connecting it with other parts of the country, to serve any industries that are developed.

    Hunters: Would like the freedom to hunt in the area.

    Present Inhabitants: Would like no changes.

    (a) Copy the map (Figure 8.50) into your notebook.
    (b) Carefully decide on the future land use of the area. You can include any number of the interests listed (plus others if you wish).
    (c) Mark your chosen plan on the map using a key to explain the symbols used.
    (d) Give an explanation of the reasons for your plan. Include reasons for your rejecting the plans of some groups.

The Fishing Industry

One of Canada's key natural resources is its fish. In 1984 Canadians caught 1 199 810 tonnes of live fish. This, however, falls short of the approximately 10 000 000 tonnes each caught by Japan and the U.S.S.R. in the same year. Nevertheless, Canadian fish exports were valued at $1 590 000 000 in 1984 and were the highest of any country in the world.

Canada controls three major commercial fishing areas. The first two are the coastal waters of the Atlantic and Pacific Oceans. The third includes the fresh water streams, lakes, and rivers of Canada's interior. Of these three areas, the Atlantic zone is the most valuable economically.

Figure 8-51 Comparative value of fishing products

THE ATLANTIC FISHERIES

Among all the great fishing areas of the world, the Grand Banks is one of the best. For hundreds of years, fishing fleets of numerous countries have looked to Canada's east coast as a valuable source of fish. As you can see in Figure 8.52, the Grand Banks is an area of shallow water just east of Nova Scotia and Newfoundland.

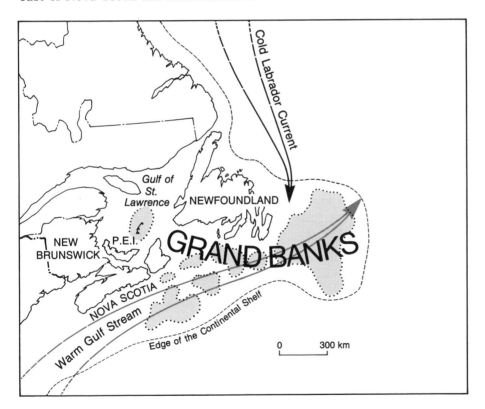

Figure 8-52 The Grand Banks *This is an area of shallow water just east of Nova Scotia and Newfoundland. Here the warm Gulf Stream meets the cold Labrador Current.*

The abundance of fish on the Grand Banks is due in large measure to the huge quantities of **plankton** living there. These are tiny plants and animals upon which fish feed.

Figure 8-53 A few varieties of plankton
The diagram illustrates the appearance of plankton when seen through a microscope.

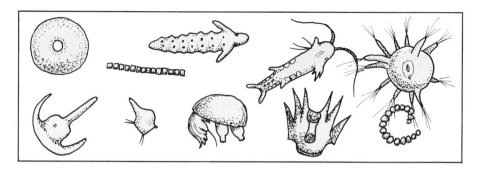

There are two major factors that allow for the abundant growth of plankton. The first is the mixing of two ocean currents — the cold Labrador Current and the warm Gulf Stream. As these two currents mix, minerals in the ocean are stirred up. It is these minerals that provide the food so essential for the growth of plankton. As the plankton multiply, so do the fish that feed upon them.

The second factor important to the fish population of the Grand Banks is the shallow water. This shallow water permits sunlight to penetrate to the ocean floor, thus encouraging plankton growth.

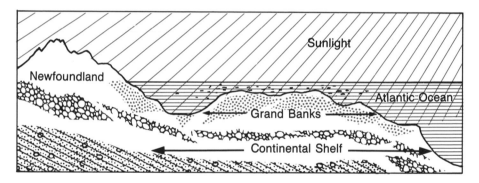

Figure 8-54 A cross section of the ocean floor showing the Grand Banks

Unfortunately, when the two ocean currents meet on the Grand Banks a great deal of fog develops. The warm air from over the Gulf Stream mixes with the cold air from over the Labrador Current to produce the foggiest conditions in the world. These fogs often cause problems for shipping.

37. (a) From a world map of ocean currents in your atlas determine the origin of the Gulf Stream and the Labrador Current. Name the body of water each current comes from.
    (b) Describe the climate of the area of origin of these currents. Use your atlas for this question also.
    (c) What would be the results for Canada's east coast and its fishing if suddenly the Labrador Current disappeared?
38. As noted earlier, the population of plankton and fish are linked. What would happen to the fish population of the Grand Banks if the amount of minerals in the water went up and then down? Explain the reasons for your answer.

FISHING OFF THE COAST OF THE ATLANTIC PROVINCES

There are two major types of fishing in the Atlantic Ocean off Canada's east coast. If you worked for yourself, you would most likely be involved in **inshore fishing.** However, if you worked for a large company, you would probably be involved in **offshore fishing.**

The table below compares the characteristics of these two types of fishing.

Figure 8-55 An inshore fishing boat

Figure 8-56 An offshore fishing vessel

Characteristics of Inshore Fishing	Characteristics of Offshore Fishing
• Small boats, limited gear.	• Large company owns large ship(s) and a wide variety of gear.
• You are your "own boss."	• You work for a company.
• Employs 85% of fishing people.	• Employs only 15% of fishing people.
• Catches 10% of total haul.	• Catches 90% of total haul.
• Fish cleaned and cured by one family.	• Fish taken to large factory in town for processing.
• Go out each day, come home at night. Stay home in bad weather.	• Go out for two weeks or more at a time. Fish in good or bad weather.
• Income low and variable.	• Income reasonable and steady.
• Independent.	• Tied to employer.

39. **Write a half page to describe the difficulties that someone involved in inshore fishing would have. Be sure to refer to offshore fishing and its impact on the fish population.**

Fishing Methods and Fish Caught

There are many different ways in which fish are caught. Much depends on whether the fish you want to catch live near the surface or near the floor of the ocean. Figure 8.57 shows you some frequently used methods.

1. PURSE SEINING
A long net with floats at the top and weights and a drawstring at the bottom is let out from the ship. A small boat pulls the net to form a circle around a school of fish. The drawstring is pulled to close the bottom of the net. The whole net is then dragged towards the ship, and the fish are scooped out with a large dip net.

Purse seining is used to catch fish that live near the surface of the water. Sardines and other types of herrings are caught this way.

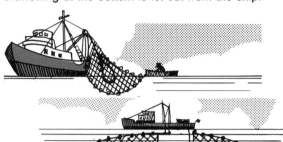
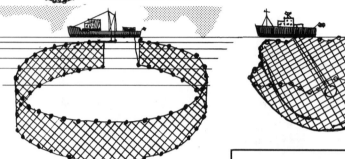

3. OTTER TRAWLING (or DRAGNET FISHING)
A bag-shaped net held open by wooden or metal "doors" is dragged along the sea floor. The fish swim in and are trapped in the narrow end. The net is hauled on board where a special knot is untied to release the fish onto the ship.

Cod, ocean perch, sole, and flounder are all caught by otter trawling. Each of these is a **groundfish,** which means that it lives near the floor of the ocean.

2. GILL NETTING
A net with floats on the top and weights on the bottom is let out from the back of the boat. The fish get tangled in the net and are then hauled in.

As with purse seining, gill netting catches fish that live near the surface of the ocean. Using this method, fishing boats catch many types of fish, including the Pacific salmon.

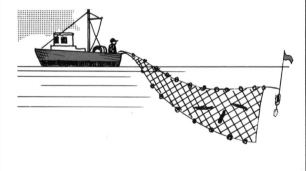
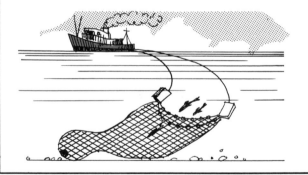

Figure 8-57 Three important fishing methods

In the Atlantic coastal fisheries the most valuable fish catches are as shown in Figure 8.58.

Figure 8-58 The most valuable types of fish caught by Canadians off the Atlantic Coast in 1985

FISH	VALUE OF CATCH ($000s)	QUANTITY CAUGHT (TONNES)
Cod	106 948	283 921
Lobster	103 277	19 119
Scallop	35 530	25 097
Queen crab	35 088	32 191
Haddock	16 325	22 299

40. (a) Using the data from Figure 8.58 determine the most valuable fish per tonne of weight. Divide the total tonnage by the value of the catch to help answer the question.
    (b) Illustrate the results of 40 (a) using a bar graph. Be sure to label each bar and the entire graph properly.
    (c) If you were to fish in the Atlantic Ocean, which fish would appear to be most profitable to catch? Base your answer only on the results of 40 (a) and 40 (b).

Canadian fishing companies are beginning to make use of factory freezer ships. On such a ship the fish are cleaned and frozen on board, rather than returning them to a land-based factory for processing.

SHELLFISH

The three main shellfish harvested in 1984 were lobsters, scallops, and crabs. Scallops live on the floor of the ocean and are caught by specially designed dredges, rakes, or tongs.

Figure 8-59 A scallop

Figure 8-60 A scallop dredge

Lobsters also live on the ocean floor, usually at a depth of 8 m to 32 m of water. Traps are specially designed to catch these lobsters. These shellfish can only move backwards, and cannot escape from the trap once they get in.

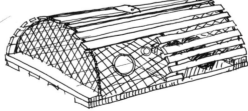

Figure 8-62 A lobster pot *Bait is put into the right-hand half of the trap. The lobster climbs into the trap through the hole, and through a second hole to the bait. It is then trapped. The lobster pot is raised to the surface and the lobster is removed through a door at the top of the trap.*

Figure 8-61 A lobster

A lobster fisherman has many lobster pots attached to lines and specially marked buoys. Every few days the traps are lifted on board using a hand operated or power-driven **winch,** which is a wheel to help raise the lobster pot. Lobsters are then removed and the pots are rebaited and replaced. Small lobsters or those which contain eggs are thrown back. The best lobsters are taken live to the market; others are canned or frozen.

---

41. Why is it important to throw some lobsters back?

THE PACIFIC COAST FISHERIES

Salmon make up about 60% of the value of fish caught in our west coast fishing areas. Herring, halibut, redfish, hake, and several kids of shellfish are other valuable catches.

Salmon are also fished off the Atlantic Provinces, but the catch is much smaller (800 t in 1984) than in the Pacific (51 000 t in 1984). As a point of interest, the Atlantic salmon are much larger, but fewer are caught. The five major species of Pacific salmon are sockeye, pink, cohoe, chinook, and chum.

The Pacific salmon have an unusual life cycle. The adult fish dies after the eggs have been fertilized. The whole life cycle is illustrated in Figure 8.63.

Figure 8-63 The life cycle of the Pacific salmon *The cycle takes several years to complete. Only one out of every 2 000 eggs hatches, becomes an adult, and returns to the spawning ground. (The Atlantic salmon does not die after spawning — it just returns to the sea.)*

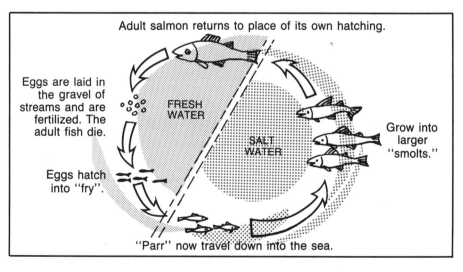

Adult salmon returns to place of its own hatching.

Eggs are laid in the gravel of streams and are fertilized. The adult fish die.

FRESH WATER

SALT WATER

Grow into larger "smolts."

Eggs hatch into "fry".

"Parr" now travel down into the sea.

The waters along the west coast provide the ideal conditions in which these salmon live. There are fast streams fed by snowmelt or heavy rain flowing into the Pacific Ocean. The salmon also find an abundance of food in the ocean, such as herring and other small fish. The fish are caught by gill nets, purse seines, or by **trolling.**

The number of salmon caught has been declining because of overfishing, pollution, and dams on the salmon-spawning rivers.

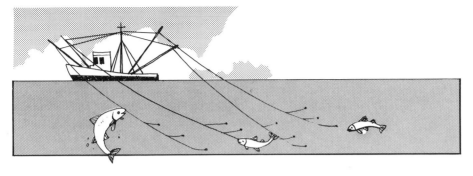

Figure 8-64 Trolling for salmon *Trolling is like sport fishing, but uses huge poles and many baited hooks. The ship moves slowly through the water. The most valuable salmon are caught by this method.*

Overfishing is now being controlled to some extent. In 1971 the Canadian government set up the **370 km fishing zone** (370 km equals 200 nautical miles). Within this distance of the Canadian coast, our government controls the type and amount of fishing that takes place. Attempts are being made to **restock** some areas. Restocking involves the introduction of young fish into a body of water in the hope that they will grow and multiply. Considerable success has resulted from a massive salmon restocking program in British Columbia.

It will require the co-operation of many people to allow the number of Pacific salmon to increase. Those who suffer from the decline of the salmon include the following:

- those who fish for the salmon
- the people who manufacture and maintain the fishing boats and gear
- those workers who process and transport the fish

Figure 8-65 Salmon — from the fishing boat to you

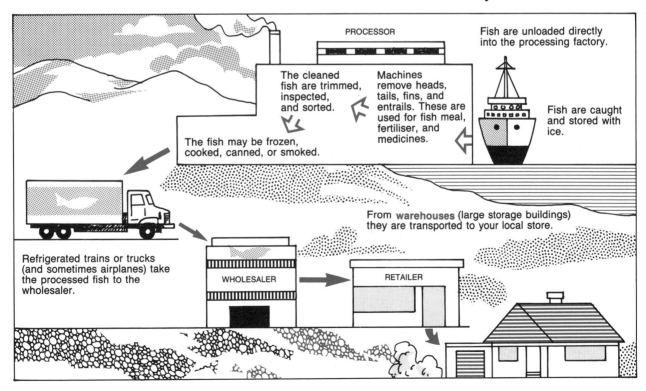

42. (a) Describe in your own words the life cycle of the Pacific salmon.
    (b) In what specific ways could human activity upset this life cycle?
43. The Atlantic fishing grounds are more productive than the Pacific grounds. Examine the maps of Grand Banks and the Pacific coast in your atlas and in this text (Figures 8.52 and 8.54 on pages 265 and 266). Explain why the Atlantic fishing areas are more productive than those in the Pacific Ocean.

THE INLAND FISHERIES

The inland fisheries are of much less importance to the economy than those of the west and east coasts. In 1984 our inland fisheries employed only 8 000 people compared to the 75 000 people engaged in ocean fishing.

44. Using an atlas map showing Canada's natural resources, list the places where most of our fresh water fish are caught. What kinds of fish are caught?

PROTECTING OUR FISHERIES

In 1971 Canada extended its fishing limits from 22 km to 370 km. The reason for this is the real concern that we have for our declining fish catches. Figure 8.66 illustrates how fish hauls have become dangerously low. They are now slowly building up again in some cases.

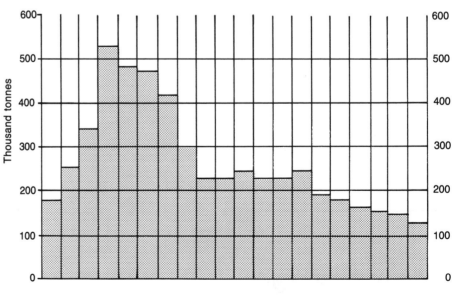

Figure 8-66 Atlantic Coast herring catch (1965-1984)

By declaring a wider fishing limit, Canada gained control over 1 000 000 km² of the Atlantic fishing area, and about 325 000 km² off the Pacific Coast. In these areas, Canadians now exercise effective conservation over the fish stocks and decide how much fish may be taken. There is also control over when, where, by whom, and with what types of gear fish are caught. The number of foreign fishing licences was cut from 1 500 to 250. Fish quotas (quantities permitted to be caught) were also reduced to allow stocks to recover. The Canadian government operates over 100 vessels off the east and west coasts to enforce these laws. Canadian Armed Forces aircraft assist in this task.

Besides the problem of overfishing, Canadian fish are also threatened by pollution and the damming of rivers. Chemicals and sewage dumped into our waters kill large numbers of fish and make others inedible for humans. In Lake Erie, for example, many fish have been contaminated with mercury, and other poisonous substances. Only careful and extensive treatment of pollutants from homes, factories, and farms can protect our fish from certain death.

Oil spills form another threat to sea life. When oil is released into lakes or oceans, it spreads out across the water surface. In doing so the oil kills thousands of fish and badly pollutes the water. Many years are needed for the water to recover from such an oil spill.

Figure 8-67 An oil-covered bird

45. Look at Figure 8.66 and describe how Atlantic herring catches have fluctuated (varied) between 1965 and 1984.
46. (a) The catch of herring from 1965 to 1984 from the Pacific is listed below. Plot these figures as a bar graph.

1965	201 425	1975	59 639
1966	139 550	1976	81 105
1967	52 953	1977	97 172
1968	2 891	1978	81 400
1969	2 003	1979	43 465
1970	3 865	1980	25 155
1971	10 017	1981	37 960
1972	39 021	1982	28 594
1973	55 625	1983	38 000
1974	44 670	1984	33 500

   (b) Describe the fluctuations in the catch of herring off the Pacific coast.
47. Compare your graph from Question 46 with Figure 8.66.
   (a) What is different about the total annual herring catch from each coast?
   (b) In what ways do the fluctuations in catch from the Pacific coast fisheries differ from those from the Atlantic coast?
48. (a) Why do fish catches give a good indication of the state of the fish stocks?

(b) What has been happening to the stocks of herring between 1965 and 1984?
(c) What might have caused these changes in herring stocks?

As discussed before, the damming of rivers can prevent fish such as salmon from swimming upstream to their spawning grounds. Sometimes fish ladders have been built to allow fish to by-pass the dam and reach the source of the river.

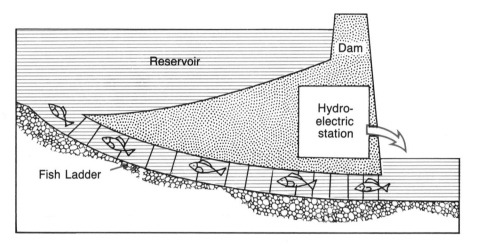

Figure 8-68 A fish ladder *"Ladders" are built around dams on salmon rivers. They are partly successful at allowing the salmon upstream. However, many young "fry" are killed in the turbines of hydro-electric stations.*

Slowly, Canadians are beginning to learn that only careful management will preserve a resource as valuable as fish for the future.

Natural resources
Renewable resources
Non-renewable resources
Exhausted
Exploited
Ore
Core samples
Mineral rights
Strip or open pit mining
Underground or shaft mining
Overburden
Shaft
Drifts
Excavated

Stopes
Pellets
Net importer
Inland sea
Prorationing
Floor price
Royalties
Fabricated
Access
Inventory
Clear cutting
Prescribed burning
Selective cutting
Skidders

Plankton
Inshore fishing
Offshore fishing
Purse seining
Gill netting
Otter trawling
   or dragnet fishing
Groundfish
Winch
Trolling
370 km fishing zone
Restock

Research Questions

1. Choose any mineral (except potash, oil, and gas) mined in Canada.
   (a) Draw a map showing where the deposits are located.
   (b) Describe how it is mined.
   (c) Find out its major uses.
   (d) Find out how valuable the mineral is as an export.
2. Choose one example illustrating that the exploitation of our natural resources has encouraged the development of our country. You might choose to study fur trading, a gold rush, the search for oil or hydro-electric power. Outline the specific changes it has brought to Canada.
3. Choose one of the products from the forestry industry shown in Figure 8.26 on page 252. Write at least one page describing the processes involved in its manufacture. Use diagrams, photographs, or samples where possible.
4. Much controversy surrounds the harvesting of seals and whales around our coast. Choose one of these animals and write a summary of the arguments for and against the hunt, using factual material.

9 CANADA'S INDUSTRIES

The Importance of Industry

Canada's wealth and prosperity are directly related to the health of its industry. Perhaps the word "industry" creates pictures in your mind of large factories with thousands of workers, pollution, heavy traffic, and noise. Such a picture is not only inaccurate; it is also narrow. Canadian industry is often very different from that.

Wherever there are people, there is industry of some description. In fact, **industry** occurs wherever a good or service is provided. The owner of an industry attempts to make a **profit.** When all expenses of an industry are paid, the profit is the money left over. This **profit motive,** or desire to make money from an industry, is essential to Canada's whole economy.

Figure 9-1 A Canadian industrial plant in an industrial park

1. Look around your classroom and record ten different manufactured items that you can see. For each item, fill in the information for the table below. Note the one example given.

MANUFACTURED ITEM	NAME OF MANUFACTURER	MATERIALS USED IN PRODUCING ITEM	TYPE(S) OF TRANSPORTATION THAT BRINGS THE ITEM FROM THE MANUFACTURER TO YOUR SCHOOL
Ballpoint	Bic	Plastic, steel, ink	Truck

2. (a) Consider the education system as an industry. List five different jobs that people might have in the educational system.
   (b) Choose one of the jobs from 2 (a). List five different ways that a person in that job might spend money to support a family. Which industries would be involved in producing each of these goods or services?

Types of Industry

As you have begun to discover, there is a wide variety of industries in Canada. Within each industry, an additional number of jobs exist. If you examine Figure 9.2 you will see that there are three different divisions of Canadian industry.

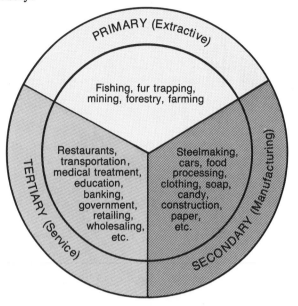

Figure 9-2 Three major categories of industry in Canada

Primary industries, as you have learned from previous chapters, form a very important part of the Canadian economy.

In recent years tertiary industries have grown most quickly of the three sectors of our economy. Primary industries have grown at the slowest rate.

Primary industries were the first ones in Canada, and still form the foundation of our economy. Although some resources extracted by primary industry are sold to the consumer, most are sent to manufacturing industries.

Figure 9-3 Our most important manufacturing centres

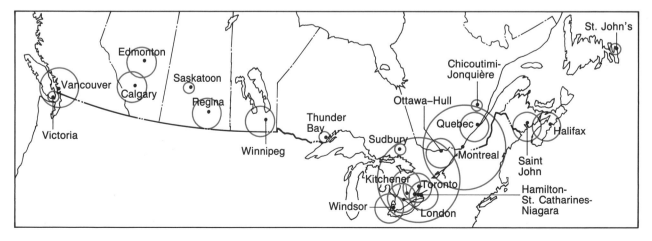

Vancouver	$ 7 567 445	Ottawa-Hull	$ 2 356 844
Victoria	372 684	Sudbury	423 184
Edmonton	4 871 248	Toronto	29 492 422
Calgary	2 590 756	Hamilton	7 301 461
Saskatchewan	571 420	Kitchener	3 424 749
Regina	715 337	London	1 778 108
Winnipeg	3 383 549	St. Catharines–	
Thunder Bay	893 713	Niagara	3 355 781
Chicoutimi–		Windsor	2 858 037
Jonquière	1 231 579	St. John's	221 375
Quebec	2 054 414	Saint John	1 785 995
Montreal	23 203 710	Halifax	1 151 265

Figure 9-4 Value of manufacturing in the 22 major centres

3. (a) Everyone employed in Canada is in primary, secondary, or tertiary industry. (For the purposes of this question, a person is in the same industry as that person's employer. For example, a cook working in a lumber camp would be involved in primary industry.) Classify each of the following jobs into primary, secondary, or tertiary industry: assembly line worker at Ford Motor Company, barber, wig salesperson, politician, hockey player, radio announcer, professional hunter, a carpenter employed in a mine, a carpenter constructing houses, a bullfighter.

   (b) When you leave school, in what category of industry would you likely get a job? Explain the reasons for your answer.

4. Figure 9.3 shows clearly where the major manufacturing centres are located and also their comparative importance.

   (a) Which two cities in Canada produce the greatest value of manufactured goods?

   (b) Write a description, from the accompanying pie graph, of the types of manufacturing that take place in Montreal. Include information about the importance of each manufacturing type relative to the others.
   NOTE: "Other goods" includes a variety of manufacturing industries.

5. Use a map in your atlas showing manufacturing in Canada to answer these questions:

   (a) Describe the types of manufacturing that occur in the capital of your province or the province closest to where you live.

   (b) Describe the difference you can observe between manufacturing in your provincial capital and Montreal.

   (c) Find out, and list, the major type of manufacturing that occurs in each of the following places:
   Vancouver, B.C.; Strathcona, Alta.; Saskatoon, Sask.; Brandon, Man.; Thunder Bay, Ont.; Sorel, Que.; Moncton, N.B.; Trenton, N.S.; Charlottetown, P.E.I.; and St. John's, Nfld.

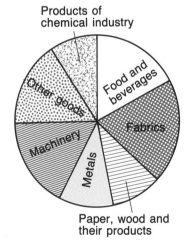

Products of chemical industry

Food and beverages

Other goods

Fabrics

Machinery

Metals

Paper, wood and their products

Factors That Affect the Location of Industries

1. *CLOSENESS TO RAW MATERIALS*

Some manufacturing industries depend heavily upon one raw material. The pulp and paper industry, for example, depends very much upon the forests for its existence. Trees, however, are very bulky and expensive to move. As a result of this fact, most pulp and paper factories are located in or near forests.

Other industries involve perishable goods such as milk, fruit, or vegetables. The processing of such items usually takes place near where they are produced. It is often the case that fish are unloaded from the ship or boat, straight into the canning plant. Such a location greatly reduces the chance of spoilage of the product.

2. *POWER*

The availability of power is a consideration of all manufacturers. For some types of manufacturing, however, vast quantities of power are needed. The manufacturing of aluminum, for example, requires more electricity than any other manufacturing process. As a result, aluminum smelters are located near a reliable but cheap electrical supply, usually from a hydro-electric power station.

At present, power is readily available to most areas of southern Canada. In fact, there are some areas of Canada with a surplus of electricity. This means that power is not as important as other factors in determining the location of an industry.

3. *LABOUR*

All manufacturing processes require workers. In or near large cities, a great number of workers are available. If development is to take place in a remote area, however, special steps must be taken to attract workers. In some cases entire towns have been carved out of the wilderness to house mine or forestry workers. High wages and special entertainment facilities are also used to attract workers and their families to remote areas. Examples of such towns include Thompson, Manitoba, Arvida, Quebec, and Terrace Bay, Ontario.

4. *TRANSPORTATION*

Another key consideration in the location of a factory is transportation. All factories depend upon good transportation to bring in raw materials and ship out finished products. If there are delays in the shipment of raw materials, a factory may have to close down its operations. Finished goods that do not leave the factory due to transportation problems may be damaged and will earn no money for the company.

6. Examine Figure 9.5 and explain why this factory is well located from the point of view of transportation.

Figure 9-5 An aerial
view of a factory

5. *MARKET*

The aim of manufacturing industries is to sell their products to the public.
The manufactured goods are most easily sold in large cities or other centres of population. In order to cut transportation costs, therefore, industries frequently locate in or close to large cities.

There are a number of other factors that may be considered by a
manufacturer when locating a factory. Some of these are explained here.

- The availability of CAPITAL (money) that can be borrowed to build
  and equip the factory is important.
- The type of LAND available can influence the location of an industry.
  Flat land with plenty of room to expand is a prime consideration for
  most companies. Services such as water supply, sewers, and roads
  should be supplied before a factory is set up.
- The other industries located in the area are important for a factory.
  These other industries help in providing supplies or markets for a new
  factory.

The manufacturing of an item adds value to the materials used to
make it. A gold ring, for example, is worth more than a plain piece of
gold. Below is a list of the value added by the ten largest manufacturing
industries in Canada (1981).

INDUSTRIAL GROUP	TOTAL VALUE ADDED	INDUSTRIAL GROUP	TOTAL VALUE ADDED
Food and beverages	9 562	Metal fabricating industries	5 929
Wood industries	3 374	Machinery industries	4 215
Paper related industries	6 944	Transportation equipment industries	7 218
Printing and publishing related industries	4 018	Electrical products industries	4 644
Primary metal industries	5 747	Chemical and chemical products industries	5 597

NOTE: Figures represent millions of dollars.

281

7. (a) Using the chart on the previous page, rank the industries in decreasing order of value added.
   (b) Plot a bar graph to illustrate this information, keeping them in ranked order.
   (c) Which of the industries depend directly upon our natural resources?
8. Refer to Figure 9.3.
   (a) Describe the location of the major manufacturing centres in Canada.
   (b) Explain the reasons for the concentration of manufacturing in the Great Lakes-St. Lawrence region of Canada referring to the location factors on pages 280-281 whenever possible.

THE IMPORTANCE OF THE GREAT LAKES-ST. LAWRENCE SEAWAY

Figure 9.3 shows that manufacturing in Canada is concentrated around the shores of the Great Lakes and along the banks of the St. Lawrence River. One very important reason for this is the ease with which raw materials and finished products can be transported in the area.

While the river and lakes have always been important routeways, it was not until 1959 that large ocean-going ships could travel from the Atlantic Ocean to Lake Superior. At a cost of $1 000 000 000, Canada and the U.S.A. had built **canals,** which are artificial rivers, to by-pass rapids and falls. They had also deepens ports and shallow areas by **dredging.** Dredging deepens the waterway by removing mud from the bottom of the river or lake. It is then put onto the land. Dredging is continually carried out to keep the water at a depth of at least 8.2 m for the large ships.

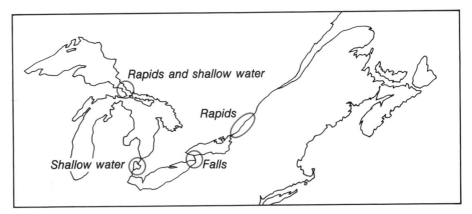

Figure 9-6 Obstacles to be overcome in the Great Lakes and St. Lawrence River

The greatest single obstruction in the system was Niagara Falls. For many years, canals had existed to allow certain kinds of craft from Lake Erie into Lake Ontario. In 1959, the new 8-lock Welland Canal was opened. One of the locks can be seen in Figure 9.7.

Figure 9-7 The Welland Canal by-passes Niagara Falls

(a) A ship in a lock on the Welland Canal

(b) A map of the Niagara area

(c) Niagara Falls

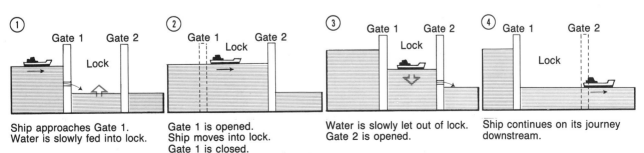

① Ship approaches Gate 1. Water is slowly fed into lock.

② Gate 1 is opened. Ship moves into lock. Gate 1 is closed.

③ Water is slowly let out of lock. Gate 2 is opened.

④ Ship continues on its journey downstream.

Figure 9-8 How a lock is used to move a ship downstream

The whole system includes three sets of canals.

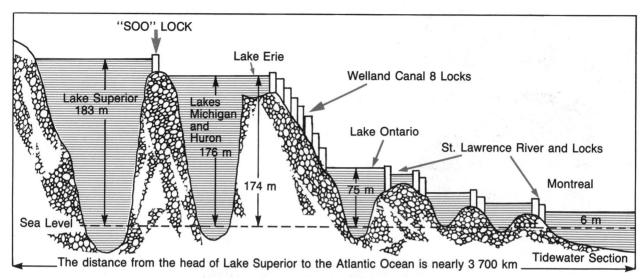

Figure 9-9 A profile of the Great Lakes–St. Lawrence Waterway

Each time that a ship passes through a canal the ship's owner has to pay a **toll.** A toll is payment for passage along a road, bridge, or waterway. In this way some of the cost of construction and operation is recovered. Another way of recovering the cost is by selling the hydro-electric power created in the International Rapids section. After the residents had been moved, very large dams in this section created huge reservoirs, which flooded over houses, roads, and railways. People involved in the movement of wheat, coal, and iron ore probably benefited the most when the waterway was completed.

9. Explain carefully why the waterway was built.
10. (a) On the waterway, almost all agricultural products move *downbound,* toward the sea. Why is this?
    (b) Most manufactured products travel *upbound,* away from the sea. Where are they coming from?
11. Why do you think that the ports of Halifax and Montreal lost some business when the waterway was opened?
12. Imagine that the Great Lakes and the St. Lawrence River disappeared. What effects would there be on landscape, climate, agriculture, transportation, industries, and cities?

THE LOCATION OF THE ALUMINUM INDUSTRY

Aluminum is a metal made from **bauxite,** a raw material which has to be imported into Canada. The other material requirements are cryolite and fluorite from Greenland and Newfoundland respectively. When these three ingredients are mixed together, an electric current is passed through them to create aluminum. Electricity is used in quantities far larger than would be available in most locations in Canada.

Figure 9-10 The aluminum smelter at Jonquière *The Grande Baie Works incorporates the most advanced technology and has three potlines in operation for an annual capacity of 171 000 tonnes.*

13. (a) Using your atlas, locate and label the following aluminum smelters on a blank map of eastern Canada: Jonquière, La Baie, Alma, Shawinigan, Beauharnois, and Bécancour.
    (b) On your map, list opposite each site which of the five major factors of location it appears to have. For example, if a city is located on a river it would appear to have a source of power. Remember as well the sources of raw materials.
    (c) Could an aluminum smelter be located near where you live? Explain your answer.

STEEL PRODUCTION IN CANADA

Iron and steel are among the most important metals in use today. Although they may appear to be very similar, there is one major difference between them. Iron tends to be more fragile and will break fairly easily under stress. Steel contains less carbon than iron and is much stronger as a result. Other chemicals can be added to steel as well to give it special qualities such as resistance to corrosion (rusting).

Iron and steel manufacturing is often called a heavy industry because it uses heavy equipment in the manufacturing process. By contrast, the manufacturing of smaller items, such as transistor radios or watches, is called light industry.

14. Choose one of the following objects or activities and list the steel products involved in
    (a) cooking a meal
    (b) a car or motorbike
    (c) building a house
    (d) a day in school.

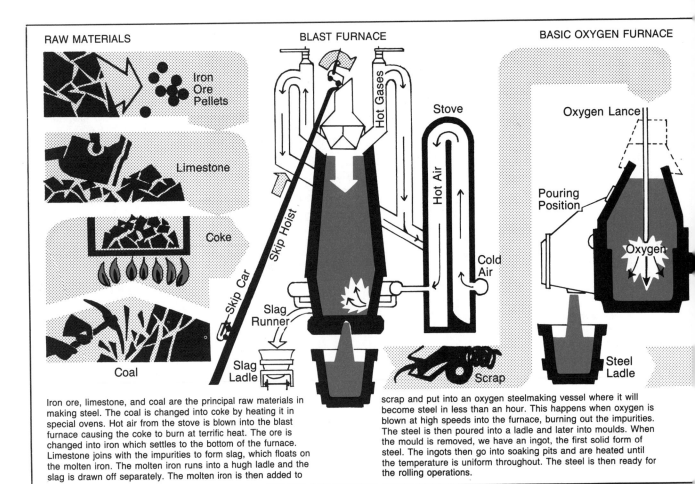

RAW MATERIALS

Iron Ore Pellets

Limestone

Coke

Coal

BLAST FURNACE

Hot Gases

Stove

Hot Air

Cold Air

Skip Hoist

Skip Car

Slag Runner

Slag Ladle

BASIC OXYGEN FURNACE

Oxygen Lance

Pouring Position

Oxygen

Scrap

Steel Ladle

Iron ore, limestone, and coal are the principal raw materials in making steel. The coal is changed into coke by heating it in special ovens. Hot air from the stove is blown into the blast furnace causing the coke to burn at terrific heat. The ore is changed into iron which settles to the bottom of the furnace. Limestone joins with the impurities to form slag, which floats on the molten iron. The molten iron runs into a hugh ladle and the slag is drawn off separately. The molten iron is then added to

scrap and put into an oxygen steelmaking vessel where it will become steel in less than an hour. This happens when oxygen is blown at high speeds into the furnace, burning out the impurities. The steel is then poured into a ladle and later into moulds. When the mould is removed, we have an ingot, the first solid form of steel. The ingots then go into soaking pits and are heated until the temperature is uniform throughout. The steel is then ready for the rolling operations.

Figure 9-11 The steel-making process

(a) A Blast Furnace *A blast furnace is tapped when sufficient liquid iron has accumulated. This happens about nine times each day.*

(b) "Teeming" Steel into Ingot Moulds *The steelmaking furnace can produce a "heat" or batch of steel in 25 minutes. The steel is poured into a ladle and then teemed into ingot moulds.*

286

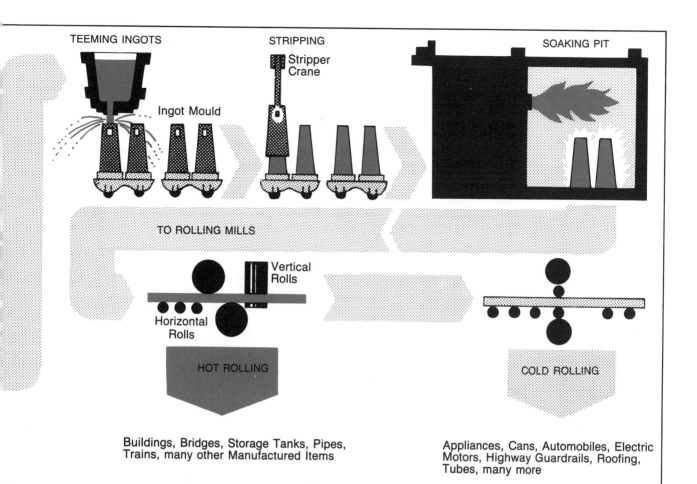

TEEMING INGOTS

STRIPPING

Stripper Crane

Ingot Mould

SOAKING PIT

TO ROLLING MILLS

Vertical Rolls

Horizontal Rolls

HOT ROLLING

Buildings, Bridges, Storage Tanks, Pipes, Trains, many other Manufactured Items

COLD ROLLING

Appliances, Cans, Automobiles, Electric Motors, Highway Guardrails, Roofing, Tubes, many more

(c) Stripping Ingots *Ingots take about 25 minutes to solidify before the moulds are stripped off.*

(d) The Rolling Mill *Each ingot is rolled into long strips at speeds of 700 m/min. It may then receive special treatment on another line.*

Mills in Hamilton, Ontario, produce over half of the steel made in Canada. The main reasons for Hamilton's importance in this industry can be traced to its location with respect to raw materials, labour, markets, and transportation.

The Raw Materials

Making steel is usually done in two stages. The first involves making molten iron from iron ore **(smelting)**. In the second stage the iron is purified further to produce steel **(refining)**.

To make one tonne of liquid iron takes approximately. . .

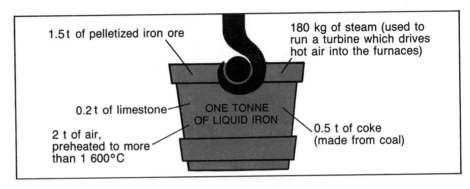

1.5 t of pelletized iron ore

180 kg of steam (used to run a turbine which drives hot air into the furnaces)

0.2 t of limestone

ONE TONNE OF LIQUID IRON

2 t of air, preheated to more than 1 600°C

0.5 t of coke (made from coal)

Figure 9-12 The sources and routes for solid raw materials coming to the Dofasco steel mills in Hamilton

The major steel mills in Hamilton are situated on flat land, much of which has been reclaimed (filled in) from the harbour to the north. Water, which is essential for steel making, is available in large quantities from Lake Ontario. Figure 9.12 shows the routes and sources for the solid raw materials for Dofasco and other steel mills in Hamilton.

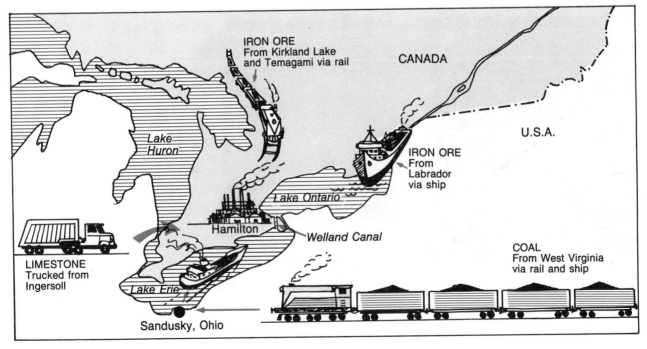

IRON ORE
From Kirkland Lake and Temagami via rail

CANADA

U.S.A.

Lake Huron

IRON ORE
From Labrador via ship

Lake Ontario

Hamilton

Welland Canal

COAL
From West Virginia via rail and ship

LIMESTONE
Trucked from Ingersoll

Lake Erie

Sandusky, Ohio

15. Copy the following table into your notes. Use information from Figure 9.12 and your atlas to help you fill in the table.

RAW MATERIAL	DISTANCE TRAVELLED TO HAMILTON (km)	METHOD(S) OF TRANSPORTATION
Iron ore from Labrador		
Iron ore from northern Ontario		
Limestone from Ingersoll, Ontario	SAMPLE	ONLY
Coal from West Virginia, U.S.A.		

16. Explain why Hamilton's steel industry benefited greatly from the opening of the Great Lakes-St. Lawrence Seaway in 1959. Your atlas will help you answer this question.

The limestone used in steel production is moved by truck to Hamilton throughout the year. Iron ore pellets also come in from northern Ontario by train all year long. The iron ore from Labrador and the coal, however, are **stockpiled** (stored in large quantities) at the steel plant before the lakes and canals freeze each winter.

Figure 9-13 Ore being unloaded by giant buckets mounted on ore bridges

The steelworks use only small amounts of Canadian coal because the mines in Nova Scotia and western Alberta are a considerable distance from Hamilton. The cost of transporting coal from Canadian sources is therefore higher than bringing it in from West Virginia.

Treating and Using Wastes

The production of steel results in many useful, as well as harmful, by-products. Some usable by-products, such as tars and oils, are collected and sold to nearby industries. Other materials, such as combustible (burnable) gases, are reused in other parts of the plant. The slag, which is drawn off from the blast furnaces, is sold to make construction blocks, fill, and insulation.

Dofasco and Stelco (major Canadian steel companies) have installed very expensive equipment to reduce pollutants.

Figure 9-14 Water pollution control plant

Water pollution is reduced by removing oils and suspended solids. Acid used in the milling operations is recycled, and water used in the cold mill is purified before it is returned to the bay.

Air pollution is reduced considerably by using scrubbers and precipitators. Figure 9.15 illustrates the effectiveness of air pollution control equipment.

Figure 9-15 Air pollution control *These photographs show the effectiveness of the air pollution control devices.*

17. List the industries in your closest city or town that use steel products.
18. Using Figure 9.13 as a guide, explain in your own words how steel is produced. Include information about by-products where it is appropriate.
19. Approximately 85% of all the steel produced in Canada comes from the following four locations:
    Hamilton, Ontario
    Sault Ste. Marie, Ontario
    Sydney, Nova Scotia
    Montreal, Quebec
    Explain the advantages of each of these places for the location of a steel industry.

Employee Relations

Encouraging co-operation between employees and management is beneficial to any industry. Dofasco provides the opportunity for its workers to become involved in plant and community activities. Money, for example, is awarded to employees who make valuable suggestions for improving steelmaking processes.

After three years with the company, employees join the profit-sharing plan. In addition to a large recreational program, Dofasco employees and their families are all invited to a mammoth annual Christmas party. The Dofasco Choir, famous for its Christmas performances on television, draws its members from within the company. Many employees have also become active in community projects (and municipal government) through their involvement with the firm.

Figure 9-16 Members of the Dofasco Choir practising for a concert

20. In what ways would the measures outlined above be beneficial to the company and its shareholders, the employees, its suppliers and customers, and the community in general?

THE AUTOMOBILE INDUSTRY

Each year, Canadians spend approximately $40 billion on cars, trucks, and their parts. There are about 15 000 000 vehicles on Canada's roads, and this creates nearly 400 000 jobs in the country.

21. In Canada, 128 000 people are directly employed in making cars, trucks, and parts. List at least 10 other types of work that depend upon the automobile industry.
22. (a) On average, transportation accounts for 13% of each Canadian's expenses. This is made up as follows:
    • 40% for buying the car
    • 22% for gasoline, oil and grease
    • 16% for parts and repairs

- 7% for insurance, parking, car washes, and driving lessons

The remainder is spent on such public transport as subways, trains, and airplanes. Construct 2 bar graphs to show the following:

(i) How much of our total expenses are for transportation. (You should have 1 divided column, or 2 separate columns.)

(ii) How our transportation costs are divided up. (You should have 5 columns.)

(b) Owning a car is the dream of many young people.

(i) What type of car do you hope to own?

(ii) List the approximate costs involved in owning this car.

(iii) At what age do you expect to be able to afford to buy the car?

The more than 60 raw materials required for making a car come from all over the world. They vary from the more obvious 1200 kg of steel and 100 kg of glass, to the unexpected coconut oil, used in paints, and beeswax, used in wire insulation.

A summary of these materials is seen in Figure 9.17. It should be noted that ceramic materials are replacing some metals.

Figure 9-17 The materials used to make a car

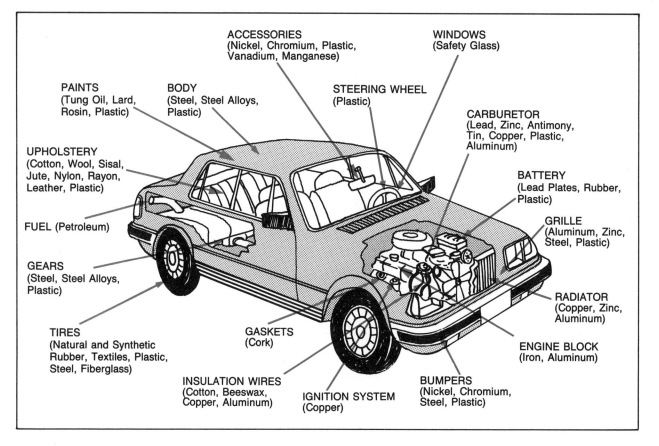

To get a new car from the design stage to the customer takes over two years.

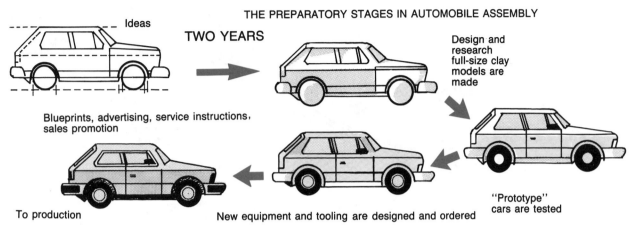

THE PREPARATORY STAGES IN AUTOMOBILE ASSEMBLY

Ideas

TWO YEARS

Design and research full-size clay models are made

Blueprints, advertising, service instructions, sales promotion

To production

New equipment and tooling are designed and ordered

"Prototype" cars are tested

Figure 9-18 The preparatory stages in automobile assembly

About 15 000 different parts are used in the production of a modern car. Many of these are purchased from other factories in Canada and in the U.S.A. General Motors, for example, depends upon the products of 35 000 suppliers. The car-assembly plant must receive them in a steady stream to keep production going smoothly and without interruption.

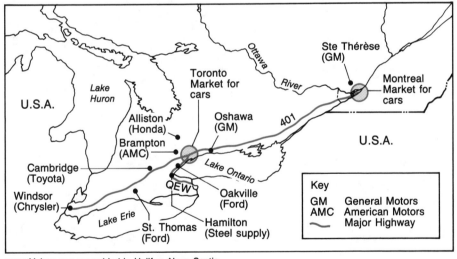

NOTE: Volvos are assembled in Halifax, Nova Scotia.

Figure 9-19 Location of automobile assembly plants in Canada

23. (a) Describe the location of Canadian automobile plants.
    (b) Explain the advantages of these locations with respect to road, rail, and air transportation. Make special note of the ease of access to the U.S.A.

The final stage in car manufacturing is by means of an **assembly line.** An assembly line is a moving conveyor that carries the item being made past teams of workers. Each team does a special part of the job.

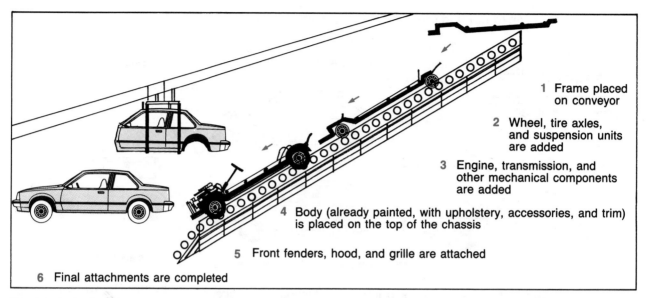

1 Frame placed on conveyor

2 Wheel, tire axles, and suspension units are added

3 Engine, transmission, and other mechanical components are added

4 Body (already painted, with upholstery, accessories, and trim) is placed on the top of the chassis

5 Front fenders, hood, and grille are attached

6 Final attachments are completed

Figure 9-20 Sequence of the car-assembly procedure

After the car is fully assembled, it is driven away under its own power. It is inspected and any necessary adjustments are made. The car is then shipped to the dealer.

Figure 9-21 One part of the car-assembly line
Robots are becoming a very important part of the assembly line.

24. What do you think are the advantages and disadvantages of assembly-line manufacture
    (a) from the point of view of the manufacturer?
    (b) from the point of view of the worker?
25. (a) What advantages are there in using robots in car assembly? List at least three.
    (b) How might this affect people who work in the car assembly plants?

SMALL BUSINESS OPERATIONS

Over 30% of businesses in Canada employ less than five people each. The people in these small enterprises make up only about 1% of the total work force in manufacturing. Small businesses, however, do form a vital section of our economy. These small enterprises are often called cottage industries, as in the example that follows.

Case Study: The Manufacturing of Stained Glass Items

In 1978, Neil Ryan of Brampton, Ontario, started a small stained glass business. The business was very successful, and in 1982, he moved to larger premises. Unfortunately, Mr. Ryan's failing eyesight forced him to sell the business to John Kerk and Mike Chesterman in 1985.

Mr. Kerk and Mr. Chesterman employ two people. One of these employees helps them to make and repair stained glass articles. The other is the store clerk who is also training to become a stained glass artist.

Income for the business comes from three main sources:
- The sale of supplies to hobbyists. This is the chief source of income.
- Mr. Kerk, Mr. Chesterman, and one of the workers teach classes in stained glass techniques. Their students then become their customers.
- They accept commissions to create or repair stained glass objects.

Their main expenses include the rent of their studio, the cost of supplies, tools and equipment, and the salaries of the two employees.

Figure 9-22 Three stained glass windows on sale in the studio

Details of the stained glass windows shown in Figure 9.22 are summarized below.

	COST OF MATERIALS	HOURS OF WORK	ASKING PRICE
Left	$50	16	$165
Centre	$80 for the copper wheel engraving bought at an auction, and $50 for the border	2	$250
Right	$50	16	$175

26. Use the information which is given about each of the windows. Subtract the cost of the materials from the asking price, and then work out the rate-of-pay per hour for each piece.
27. Why is crafting items with stained glass more suitable for a small cottage industry than a large manufacturing plant?
28. List, and briefly describe, any small industries you are aware of in your local area.
29. Which of the activities you are interested in could be developed into a small business? Explain how you would develop your own establishment.

RESTAURANTS—A TERTIARY INDUSTRY

The restaurant industry is familiar to almost all Canadians. More and more people eat away from home, whether it is in a very expensive restaurant, a fast food outlet, or a high-school cafeteria.

Figure 9.23 lists the restaurant organizations with the highest annual sales of food, beverages, and lodging in 1981 and 1984 in Canada.

Figure 9-23 The top ten in food services

ORGANIZATIONS	1981	
	SALES ($ MILLION)	OUTLETS
McDonald's	600	450
Cara Operations	323.6	—
C.P. Hotels	196.3	100
Scott's Hospitality	158.6	—
V.S. Services	143	500
Dept. of National Defence	134.9	200
Les Patisseries-St. Hubert	110	68
Arvak Management	109	504
Keg Restaurants	108.2	80
Four Seasons Hotels	105.2	—

ORGANIZATIONS	1984	
	SALES ($ MILLION)	OUTLETS
McDonald's	880	464
Cara Operations	445	377
Scott's Hospitality	319	523
V.S. Services	228	2600
Four Seasons Hotels	171	—
Dept. of National Defence	158	200
C.P. Hotels	152	—
Keg Restaurants	150	105
Burger King	149	127
Les Rotisseries-St. Hubert	144	96

Sales figures (Figure 9.23) represent gross (total) annual receipts from food and beverages. In 1984, fast-food outlets accounted for almost 40% of the total $10.8 billion of total sales from restaurants in Canada. The value of fast food service from convenience stores makes up 10% of their sales.

---

30. Here is a list of the top ten food items ordered in 1984.

French fries	24.8
Salad	16.4
Bread/toast/muffin/bagel	14.9
Hamburger	14.4
Soup	11.6
Sandwich	10.6
Cake/pie, pastry, etc.	10.3
Vegetables	9.3
Nothing	8.6
Other — potatoes	8.5

These figures are expressed as percentages. They add up to more than 100 because many people order more than one item.

(a) Explain what might have been consumed by the people in the "nothing" category.

(b) Construct a graph to illustrate the figures shown above. Your graph should be large, colourful, and imaginative. It could be a bar or pie graph. Use appropriate symbols to show the different food types, and include a title and a key if necessary.

31. (a) In your school cafeteria, and for a period of five minutes, note the purchases of students as they exit the food line. To make this task easier, get a piece of paper and write each of the above categories as a list down the left hand side. Add your own categories such as "ice cream" if you think that there is a big demand for these items. While you are doing your survey, you only need to make a mark for each item purchased.

(b) Plot your results in a graph.

(c) What differences, if any, do you observe when you compare food preferences to those listed at the beginning of question 30?

(d) List as many reasons as you can for the differences listed in (c).

32. Referring to Figure 9.23, which organizations entered and which ones left the top ten list for food services between 1981 and 1984?

33. (a) By how much did McDonald's sales increase between 1981 and 1984?

(b) How does their increase in sales compare with that of other food chains listed?

Case Study: McDonald's Restaurants

McDonald's of Canada is just one part of a multi-national organization with its headquarters in the U.S.A. In 1986, McDonald's had outlets in the following countries.

Australia	Guatemala	Panama
Andorra	Hong Kong	Philippines
Austria	Ireland	Singapore
Belgium	Italy	Spain
Brazil	Japan	Sweden
Costa Rica	Malaysia	Switzerland
Denmark	Netherlands	Taiwan
El Salvador	Netherlands Antilles	Bahamas
England	New Zealand	Guam
France	Nicaragua	Puerto Rico
Germany	Norway	Virgin Islands

McDonald's also has a floating restaurant which was originally constructed for use at Expo 86 in Vancouver.

McDonald's of Canada is one of the foreign-owned subsidiaries (branch) of the U.S.A. corporation. The McDonald's chain began with one restaurant founded by Ray Kroc in Chicago in 1955. There are now over 8 000 outlets, and 100 shares worth $2 250 (U.S.) in 1965 today would bring the owner over $120 000 (U.S.).

To become so successful, these restaurants must offer the type of service demanded by the public. According to the company, this includes a well-lit, clean building, with a uniform but varied menu. The food must also be served quickly and at reasonable cost.

Figure 9-24 McDonald's around the world

The photographs are from England, Bahamas, Hong Kong, Austria, Japan, and Australia. Can you guess where each photograph was taken?

Many of the outlets operate under a **franchise** system. A person who buys a franchise is allowed to use a McDonald's building and the name and recipes for food preparation. Only 10% of all applicants for franchises are accepted by McDonald's. Once a person has been accepted by McDonald's two conditions must be met:
- An investment of $500 000 for the franchise must be paid to McDonald's in the United States.
- An intensive training program for that person at McDonald's Hamburger University in Illinois.

Figure 9-25 A McDonald's restaurant in Canada *McDonald's restaurants often sponsor local hockey teams.*

4% is spent on *advertising.*

11.5% *is sent back to the* U.S.A. for the use of the buildings.

10-15% *profit* is generated. The licencee would probably take 2.5% as his share.

The rest pays for *costs of* food, supplies, electricity, wages, etc.

Figure 9-26 Where the money from McDonald's goes

34. In 1984 an average store in Canada sold $1 900 000 worth of food and beverages. Using the information from Figure 9.25, calculate for each restaurant
    (a) the amount of money sent each year to the U.S.A. headquarters for use of the McDonald's building
    (b) the amount of money spent on advertising
    (c) the income of the person who owns the franchise for a McDonald's restaurant.

The supplies required for the restaurant are bought locally, although several McDonald's stores may buy from the same place. In this way they support local industry. Each outlet, on average, employs 107 workers. Many of these workers are high school students and work only part time. Each licencee is expected to spend a considerable amount of money on community activities. These activities have included field trips for orphans, help for charitable organizations to raise money, and gifts of emergency-relief food supplies. They also help senior citizens and sponsor the muscular dystrophy Labour Day telethon.

McDonald's has established several Ronald McDonald houses. In these houses, the families of children being treated for cancer, leukemia, and other serious illnesses can reside while the child receives treatment in a nearby hospital. Ronald McDonald houses exist in Toronto, Ottawa, Montreal, Halifax, Winnipeg, and Vancouver. Others are planned for Calgary, Edmonton, London, and Saskatoon. There are over 60 Ronald McDonald houses in the U.S.A.

35. List McDonald's contributions to the community, as outlined above. You may also include any projects sponsored by McDonald's that you have observed in your local area.
36. (a) With this type of foreign-owned business, in what specific ways does Canada lose?
    (b) In your opinion, do the benefits derived from the McDonald's restaurants in a region justify the money sent to the U.S.A.?

Industry
Profit
Profit motive
Primary industry
Secondary industry
Tertiary industry
Canals

Dredging
Toll
Downbound
Upbound
Bauxite
Heavy industry
Light industry

Smelting
Refining
Stockpiled
Assembly line
Cottage industries
Franchise

Research Questions

1. Kitimat, B.C., is the home of the second largest aluminum smelter in Canada.
   (a) Locate Kitimat on a map of British Columbia.
   (b) Using the map and an illustrated written account, describe how the necessary electric power for the aluminum plant was created and transmitted to Kitimat.
   (c) Explain the advantages of Kitimat for the location of an aluminum smelter.
2. There are three main ways of making steel from iron. The basic oxygen-furnace method has been described.
   (a) Find out about the other two methods of steel-making used in Canada. Using diagrams, describe them.
   (b) List the advantages and disadvantages of each of the three methods of steel manufacture.
3. Find a cottage industry in your area, interview the owner, obtain and report the following information:
   • the nature of the product(s)
   • the nature and sources of raw materials
   • the number of employees
   • the history of the development of the industry
4. Choose any manufactured product in a local store. Find out all that you can about it. It may be necessary to ask the store owner for help or to write to the manufacturer. Your account should include the following:
   • the name, nature, and composition of the object
   • its place of manufacture
   • the origin of the raw materials
5. Take part in a local industrial survey. On a map of your area supplied by your teacher, and working in groups of about four, mark on and name the various industries. Classify the industries into primary, secondary, and tertiary using a colour scheme and corresponding key. Write a summary of the industries found in your area of study.

10 CANADA AND ITS REGIONS

Canada's Political Divisions

In Chapter 1 we divided Canada into a number of physical regions. Each one of these regions was distinctive because of the nature of the landforms within it. Canada can also be divided into regions on a political basis, as in Figure 10.1.

Figure 10-1 A map of Canada, with its political regions and capitals

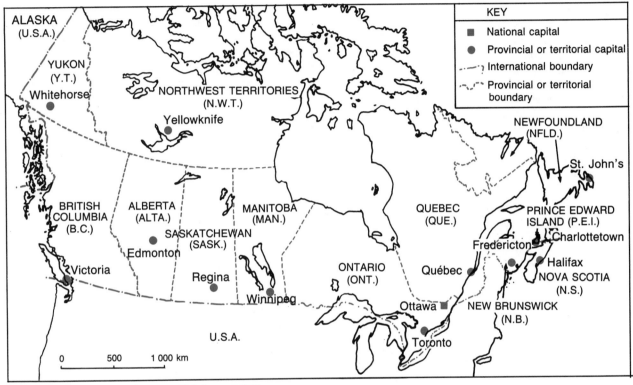

NOTE: These political units can be grouped into larger regions.

The reasons behind such a division can be summarized in the following way:

NAME OF REGION	PROVINCE(S) OR TERRITORIES INCLUDED IN EACH REGION	MAJOR FEATURES COMMON THROUGHOUT THE REGIONS
Atlantic Provinces	Newfoundland, Nova Scotia, New Brunswick, Prince Edward Island	Coastlines along Atlantic Ocean. Mostly forested. Rugged, rocky landscape.
Quebec	Quebec	French is the predominant language. Mostly heavily forested.
Ontario	Ontario	Highest population of any province. Most industrialized of all provinces.
Prairies	Manitoba, Saskatchewan, Alberta	Largely flat land. Southern part is grassland.
British Columbia	British Columbia	Mainly mountains. Very rugged, forested landscapes.
The Yukon and the Northwest Territories	Yukon Territory, Mackenzie, Keewatin, and Franklin Districts	Generally cold climate. Few inhabitants.

Those cities marked (•) on the map are the capitals of Canada's provinces and territories. In these capital cities, the provincial or territorial legislatures meet. Each legislature is composed of elected politicians who run the government. Many of the **civil servants** (government employees) live and work in the capital city as well. The largest city in the province may or may not be the capital.

Figure 10-2 Canada since 1867 *Canada has become a self-governing dominion within the Commonwealth of Nations. The process took just over 80 years. Changes in the status of the two territories will probably occur in the future, giving them more self-government.*

1. Select the capital city of your province or territory. Describe its location relative to the entire province or territory. Is it well located to serve the people of the province or territory? Explain your answer.

Canada's Political Growth

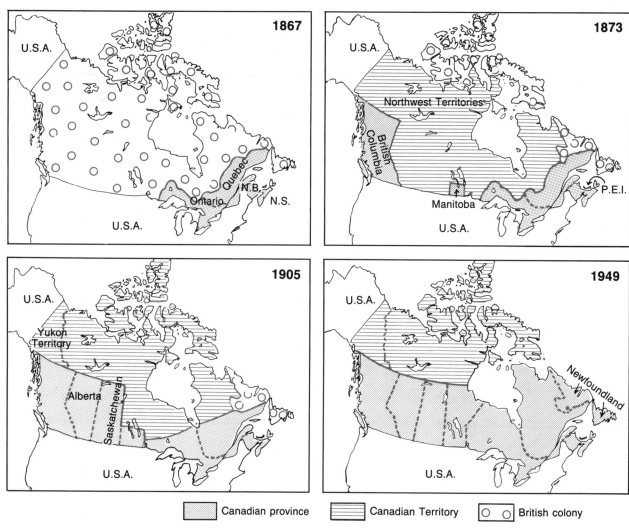

2. Describe in your own words the changes in the size and shape of Canada since it was established as a separate country in 1867.
3. (a) When did your local area become part of Canada?
   (b) What changes, if any, have occurred in your home province (territory) since it joined Canada?
4. Examine the earliest map of Canada shown. Where did the settlement occur at that time? Suggest reasons to explain this pattern of early settlement.
5. Ottawa was chosen by Queen Victoria as the site for Canada's capital, which it became in 1858. Suggest reasons to explain why she would have chosen Ottawa rather than another city already established. Make reference to Figure 10.2 in your answer.

Even in 1867, Canada had a number of strong, individual regions with their own interests and ways of life. It therefore seemed important that these people should have a voice in the government of their own area.

As a result, the **BNA (British North America) Act** was introduced to give Canada three basic levels of government. The largest government was the **federal (national) government,** which was to handle relations with other countries and problems of national importance. The provincial governments handled more local matters. Local or municipal governments were designed to run urban centres, townships, etc.

The BNA Act was a law that set up Canada as a separate nation. Besides establishing the various levels of government in Canada, the BNA Act also recognized both English and French as official languages in Canada. Canada gained the right to alter its own **constitution** in 1982. The constitution of a country is a document which outlines how the nation is to be organized and governed.

To understand Canada more fully, it is important to discover some of the characteristics of each region within our country. The questions that follow will help to alert you to the nature of Canada's regions.

Regional Studies of Canada

Earlier chapters in this book and your atlas will provide much of the information you will need to answer the questions. Some questions, however, will involve the use of other books, available from your school or local library.

6. Choose one or more regions from the list on page 303 and carry out the instructions.
   (a) On a blank map of Canada, colour in and label the region you have chosen. Give your map an appropriate title.

(b) Write a half-page account that describes the location of your chosen region. Your account should refer to Canada as a whole, the U.S.A. (where appropriate), and nearby provinces and territories.

(c) Of what vital importance to the rest of Canada is the location of this region?

7. (a) On a large map of the region, mark the major physical features. These should include mountains, hills, plateaus, rivers, lakes, and other bodies of water. Give your map a suitable title.

(b) (i) Describe the location of each major physical feature in your region. Describe any patterns of distribution that you can see.

(ii) If there is a coastline, describe its length and suitability for harbours. In addition, list countries which could trade with ports on the coast.

8. (a) Referring to Chapter 5 and your atlas, describe the types of agricultural activities that take place in the region you are studying.

(b) Explain the advantages or disadvantages the region has for agriculture.

9. List the natural resources (including hydro-electric power) that have been developed in your region of study. Mark on a map of the region the areas where each is found.

10. (a) Compared to the rest of Canada, how important is manufacturing in your area of study? What kinds of products are made?

(b) Write an account of one important industry in the region that has not yet been studied in this book. You will need to use your library to answer this question. Centre your account on the following:
- location
- specific products that are manufactured
- the methods used to manufacture the final products
- market(s) for the manufactured goods

11. What resources or products of the manufacturing industry in the region you are studying are important to Canada as a whole? Explain the reasons for your answer.

Regional Differences

Each region of Canada has its own set of problems and prospects. From time to time the interests of one region come into conflict with those of other regions. In the field of energy, for example, there are considerable differences of opinion. Since the Prairies produce most of Canada's oil, it is in their interest to receive high oil prices. In Ontario and Quebec a great deal of oil is used to run the large manufacturing industries. The residents there want to keep the oil prices low. This type of conflict works to pull Canada apart.

There are a number of other conflicts that continue to threaten the unity of Canada as a nation. A number of these conflicts are discussed here.

ECONOMIC DIFFERENCES

Although Canadians as a whole are among the richest people in the world, there are significant differences in wealth within Canada. Examine the following table.

REGION OF CANADA	AVERAGE PERSONAL INCOME	UNEMPLOYMENT (AS A PERCENTAGE OF THE WORK FORCE)
Atlantic Provinces	15 000	14.0
Quebec	14 500	8.6
Ontario	17 000	8.2
Prairies	11 500	12.0
British Columbia	15 600	16.0

Figure 10-3 Disparity in income and employment in Canada (1985)

12. (a) Draw two bar graphs to illustrate the two sets of information in Figure 10.3.
    (b) Which region of Canada appears to have the healthiest economy? Explain your answer.
13. What other statistics besides unemployment and income can you think of that would show economic differences within Canada?

In addition to disparities in income and employment, conflicts and misunderstandings can arise from the wide variety of jobs in Canada. A person involved in fishing in Newfoundland, for example, has a very different set of concerns and attitudes than an executive of a large paper company in Toronto. Each of these people expects different benefits from Canada and its government.

CONCERNS OF A LOGGER IN BRITISH COLUMBIA	CONCERNS OF AN ENVIRONMENTALIST IN MONCTON, NEW BRUNSWICK
• lower taxes	• wants higher income from the government
• higher price for paper and thus increased income	• low price for paper to keep cost lower
• physical fitness	• environmental concern
• suitable weather for logging	• more land for parks and recreation
• more land for logging	• family welfare and activities
• family welfare and activities	

14. Imagine that you are a logger in British Columbia. In what specific ways would your views of Canada differ from those of an environmentalist? Write a letter to the environmentalist to explain your point of view.

The federal government has made numerous attempts to try to reduce the economic differences within Canada. These efforts have achieved partial success. One such program is that of tax benefits for industries that are set up in poor areas. Under this arrangement, the government decides where jobs are most needed. If a company establishes a factory to provide jobs, it is rewarded by paying less tax. In the Maritimes, for example, a Volvo car-assembly plant and Michelin tire factories have been built to take advantage of this program.

Figure 10-4 A Michelin tire plant established in the Maritime provinces with government assistance

The federal government has also set up a program of **equalization payments** to the provinces. These payments take tax money from the wealthier provinces and give it to the poorer provinces. This money would be used by the provinces receiving it to improve the quality of life, work, and recreation for their residents.

FRENCH-ENGLISH DIFFERENCES

The greatest concentration of French-speaking people in North America is in the province of Quebec, where approximately 81% of the population speaks French as a **mother tongue** (language of the home). Many French people in Quebec are concerned because they are surrounded by so many English-speaking people in the United States and Canada. They believe that their special traditions, culture, and language may be eliminated unless they take action now.

Since the early 1960s there has been considerable discontent in Quebec. Some believe that the only way to maintain their French traditions is for the province of Quebec to become a separate nation. This philosophy is referred to as **separatism.**

The federal government has passed a series of laws (Figure 10.5) in an attempt to meet the needs of the **Québécois** (French-speaking residents of Quebec). Whether the complaints of the Québécois will be fully met by these federal government programs will only be seen in the future.

---

15. Look at Figure 10.5. Imagine a Canada without Quebec. Which region of Canada would be most affected? Explain the reasons for your answer fully.

16. Examine the thematic maps of Canada in this text and in your atlas. In what ways would Canada suffer economically as a result of Quebec's separation?

Figure 10-5 Problems of the Québécois, and government attempts to alleviate them

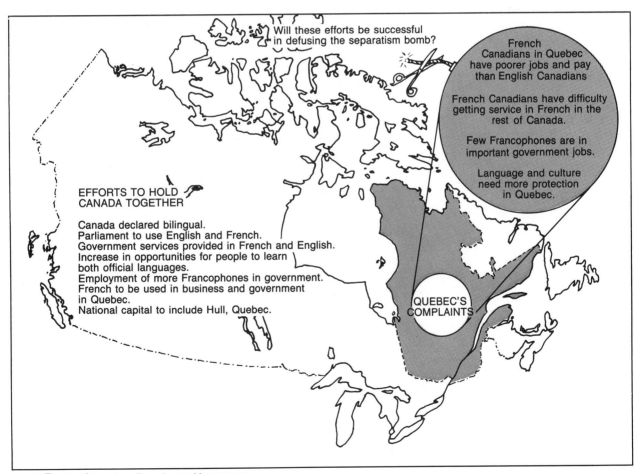

Will these efforts be successful in defusing the separatism bomb?

French Canadians in Quebec have poorer jobs and pay than English Canadians

French Canadians have difficulty getting service in French in the rest of Canada.

Few Francophones are in important government jobs.

Language and culture need more protection in Quebec.

EFFORTS TO HOLD CANADA TOGETHER

Canada declared bilingual.
Parliament to use English and French.
Government services provided in French and English.
Increase in opportunities for people to learn both official languages.
Employment of more Francophones in government.
French to be used in business and government in Quebec.
National capital to include Hull, Quebec.

QUEBEC'S COMPLAINTS

NOTE: A **Francophone** is a French-speaking person.

309

CANADA'S PATTERN OF PHYSICAL FEATURES

Canada's physical features have been an important influence on its development.

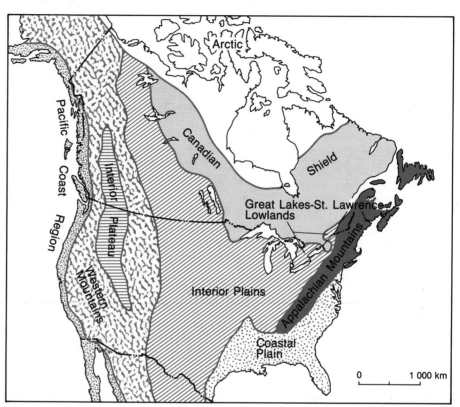

Figure 10-6 The main physical divisions of North America

Examine Figure 10.6 and answer the following questions.

17. (a) List the various physical regions in Canada and explain the direction of orientation of each of them. For example, the west coast region of British Columbia is oriented, or runs, in a north-south direciton.
    (b) Based on the results of 17 (a), if countries were set up in North America on the basis of *physical features only*, in which direction would such countries be oriented?
    (c) Assume that the boundaries of Canada and the United States were dissolved. On a blank base map of North America, draw in the boundaries of brand new countries of your own design. Base these countries on the results of 17 (a) and 17 (b).
    (d) (i) What advantages would result from your arrangement of countries?
       (ii) What disadvantages might be created?
18. What problems does Canada face in attempting to remain unified as a nation? Base your answer on your conclusions from Question 17.

Throughout its history, Canada has been very strongly influenced by the United States. For example, a great deal of trade from Canada's east and west coast ports is with the United States. Southern Ontario and Quebec are tied economically to the regions of the United States immediately south of the Great Lakes and the St. Lawrence River. In many other ways Canada is linked to the United States, as you have already discovered. The United States has a population approximately ten times the size of Canada's. The U.S.A. is also the greatest industrial power in the world, with many of the world's largest corporations and organizations. Since Canada is located on the doorstep of the United States, it sometimes appears as if Canada is sleeping next to an elephant. Everytime the elephant moves or sneezes, Canada must beware.

THE IMPORTANCE OF COMMUNICATIONS TO CANADA AND ITS PEOPLE

Increasingly, Canadians depend upon the mass media for entertainment and for news of the world around them. Recent surveys show that more people learn about daily news from the television than from any other source. Television has become very important to the lives of most Canadians, whether the programs originate in the United States or Canada.

Approximately 61% of Canadian households with television have **cable television** service. With this service one large antenna picks up television signals, which are then sent to individual homes via cable. Cable television is convenient and provides a range of programs that a regular television antenna could not receive. Cable television now also brings American television broadcasting to many Canadians who are not close to the border of the United States.

Canadians have shown a strong preference for television shows and popular music from the United States. Consistently, American television shows rate among the most popular ones shown in Canada. Many Canadians have become concerned about this strong American influence on television and radio. As a result, the CRTC (Canadian Radio and Television Commission) has set up rules to control the amount of American programming on Canadian media. Radio stations are now required to have over 30% of their music between 06:00 and 24:00 h to be of Canadian origin. Similarly, at least 50% of the programs carried by Canadian television stations during the same time period must be produced in Canada.

The **CBC** (Canadian Broadcasting Corporation) is a government-owned radio and television network designed to help unify Canada. About 80% of the CBC's content is of Canadian origin. The CBC is now aiming for 100% Canadian content, and hopes to cover a wider broadcasting area. Many performers who might have emigrated to the United States have remained in Canada due to the encouragement of the CBC. In its own way, CBC is attempting to link Canada on an east-west basis.

Figure 10-7 Behind the scenes with the Canadian Broadcasting Corporation

The federal government has also launched a series of space satellites to aid in communication. *Anik* is a new Canadian telecommunications satellite. Areas of our country that were never able to receive television or radio broadcasts before can now do so by means of Anik and other satellites.

---

19. (a) Set up a survey of your own class. Ask ten students to list their ten favourite television shows.
    (b) Once you have completed your interviews, rank the shows in order of popularity. The first show, for example, would have been mentioned by the greatest number of students.
    (c) Having completed your ranking of shows, record the country where each show originated. What percentage of the most popular shows for your class came from the United States? What percentage came from Canada?
    (d) What does this survey tell you about the influence of the United States on Canadians?
20. Imagine that all American television programs shown in Canada were cancelled. What impact would this have
    (a) on Canadian television viewers?
    (b) on the Canadian television industry?

Canadians and their government have made numerous attempts to draw together their large and diverse country. Some of these attempts have already been described. However, there are still rumblings of separatism from people in various regions of Canada.

Communications provide a vital link in holding Canada together. In almost every Canadian home there is evidence of different types of communication—from mass media such as television, radio, and newspapers to personal forms of communication.

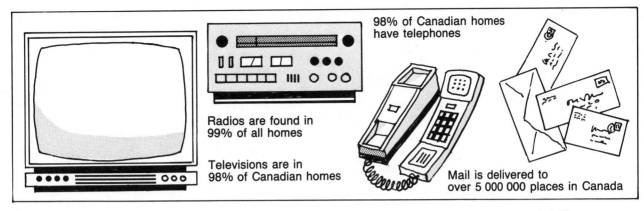

Radios are found in 99% of all homes

Televisions are in 98% of Canadian homes

98% of Canadian homes have telephones

Mail is delivered to over 5 000 000 places in Canada

Figure 10-8 How Canadians communicate

SHOULD CANADA JOIN THE UNITED STATES?

Due to the strong influence of the United States in Canada, some Canadians believe that Canada should join the United States.

ADVANTAGES OF CANADA JOINING THE UNITED STATES	DISADVANTAGES OF CANADA JOINING THE UNITES STATES
• Canada would be part of the most powerful country in the world and would have a stronger influence over other countries. • Larger market for Canadian goods. • Certain goods would be cheaper for Canadians to buy (e.g. appliances, cars, etc.) • Greater freedom for Canadians to travel, since there would be no border to cross to the United States. • As more Canadian resources are developed, more Canadians would receive jobs.	• Canada would lose its form of government, traditions, and ethnic variety. • Quebec would likely lose its culture and language. • Canada would inherit a number of American problems (e.g. crime, racial tensions, etc.) • Less control for Canadians over their own affairs. • Less careful development of our natural resources, especially in the North.

21. After examining the advantages and disadvantages of Canada joining the United States, make your own decision on the issue. Write a paragraph to explain the reasons for your answer. Include your own reasoning.

As Canadians look to the future of their country, they must decide what role Canada will play in the world. What does Canada have that is unique or distinctive? What does it mean to be a Canadian?

22. Assume that you are on a vacation and you meet a person from China. You start a conversation with that Chinese traveller and you are asked what it is like to live in Canada. Based on what you have studied in this book, explain what you believe it means to be a Canadian.

23. Create a collage (a picture made of many cut-out pictures) that depicts Canada as you see it. Write one paragraph to explain the reasons for the choices of pictures in your collage.

Vocabulary

Civil servants
BNA Act
Federal government
Constitution

Equalization payment
Mother tongue
Separatism
Québécois

Francophone
Cable television
CBC

GLOSSARY

TERM	MEANING	EXAMPLE, ILLUSTRATION, OR APPLICATION
Access	The ability to reach.	Access to the Arctic islands is difficult.
Air Cargo	Transporting goods by air.	Fresh vegetables and mail are delivered to Arctic settlements using air cargo.
Altitude	The height of a point above sea level.	The Ottawa airport has an altitude of 126 m.
Assembly line	A moving conveyor that passes teams of workers. Each worker does a special job on the passing items.	Car parts are fitted together on an assembly line.
Atmosphere	The layer of air that surrounds the earth.	
Base Line	A road used as the basis for surveying a township in the Ontario survey system.	See CONCESSION.
Bauxite	The ore from which aluminum is made.	Bauxite is not mined in Canada; it has to be imported.
Biomass Conversion	The production of usable forms of energy from plant material.	Biomass conversion may become very important in the future.
Black Soils	Rich, deep grassland soils found on the Canadian Prairies.	
BNA Act	The British North America Act.	An Act of Parliament that gave Canada three levels of government.
Boom Town	An urban centre that has grown very quickly from a small community.	Fort McMurray is a boom town in Northern Alberta where oil sands are being mined.
Break-of-Bulk Point	A settlement where a commodity is transferred from one type of transportation to another.	
Cable Television	A service whereby one central antenna picks up television signals, which are then sent to many homes via cable.	Cable television allows us to receive more channels, and to have better reception, than a household antenna.
Canadian Citizenship	A person whose nationality is officially Canadian.	Canadian citizenship is considered a valuable asset by most immigrants to Canada.
Canadian Constitution	The law that set up Canada as a separate nation.	The Canadian Constitution contains the written guidelines by which our country is governed.
Canal	A man-made river.	Canals are built so that ships may pass obstructions such as falls and rapids.

Candu Reactor	Canada Deuterium Uranium Reactor (A Canadian nuclear reactor using deuterium oxide (heavy water) and uranium).	CANDU reactors have been sold to many nations.
Capital	Money used to invest.	Before building a skyscraper, enough capital must be obtained to start the job.
Cardinal Points	The major points of a compass.	North, south, east, and west are cardinal points that will help you locate yourself anywhere in Canada.
Cartographer	A person involved in making maps.	Today, cartographers use modern air photographs and computers in making some maps.
CBC	The Canadian Broadcasting Corporation.	The CBC has radio and television stations in almost all parts of the country.
Census	A counting of the population.	Each household has to complete a census form every ten years.
Central Business District (CBD)	The downtown area of a city, usually with the tallest office buildings.	The CBD of a city often contains major shopping areas.
City	A settlement with a population of at least 10 000 people.	Cities are important economic centres in Canada.
Civil Servants	Government employees.	Many of the people who live in Ottawa are civil servants.
Clear Cutting	All usable trees are removed from an area.	Clear cutting on steep slopes may lead to soil erosion.
Climate	The average of weather conditions for a certain place over many years.	The climate for Resolute, NWT, is cold throughout the year.
Climograph	A graph showing the average temperatures and amounts of precipitation for a place during a year.	Climograph for Charlottetown, P.E.I.
Commercial Land Use	The area of land where people shop and conduct their business.	Modern shopping plazas are an important type of commercial land use.
Commercial Strip	Stores located along both sides of a major street.	
Commodities	Goods.	A wide variety of commodities are shipped across Canada by mail.
Concession	A strip of land in a township in Ontario.	

316

Concession Line	A road that runs along the edge of a concession.	See CONCESSION.
Condense	The change of state of water vapour into liquid water due to cooling.	When water vapour rises and cools, it condenses to form clouds.
Confluence	Where a tributary joins another stream or river.	Winnipeg is at the confluence of the Red and Assiniboine Rivers.
Conifers	Evergreen trees with needles rather than flat leaves.	Conifers are used in the production of pulp and paper products.
Containerized Freight	Commodities are loaded into a large container where they are produced, and then moved to the users without being unpacked.	Containerized freight saves a great deal of time in loading and unloading.
Continental Resource Program	Plans for the development of resources in the whole continent.	Continental resource programs may be beneficial to many countries.
Continental Shelves	The extension of a continent beneath shallow water.	Continental shelves often contain valuable oil deposits.
Contour Interval	The change in elevation between two contour lines found next to each other.	300 400 500 The Contour interval is 100 m
Contour Line	A line on a map that joins points of equal elevation above sea level.	275 250 225 200
Convectional Precipitation	Precipitation caused by the rising of hot air and the resulting formation of clouds.	Convectional Precipitation. Hot Air Rises
Co-Ordinates	A military grid co-ordinate is a three-figure number used for locating a place on a map. Two co-ordinates, totalling six digits, locate a place on a map.	81 82 83 84. Co-ordinates for A are 835215
Core Samples	A cylindrical sample of rock brought up by a drill.	Core samples are used to find out whether rocks buried beneath the earth's surface contain oil.
Cottage Industry	A factory that employs very few people.	Wood carving is a famous cottage industry in Quebec.
Cracking	The process of mixing a catalyst with crude oil, in preparation for dividing it into different petroleum products.	Cracking is an important stage in refining oil.
Cultural Baggage	The traditions, customs, etc. that immigrants bring with them from their homeland.	Pizza is one example of cultural baggage brought to Canada from Italy.
Cyclonic Precipitation	Precipitation that occurs when cold and warm air meet.	During the spring and fall, cyclonic precipitation occurs frequently in Canada.

Deciduous Tree	A tree with flat leaves that fall during the autumn.	Many tourists travel to Canada to see the fall colours of the leaves of our deciduous trees.
Delta	The mouth of a river that is divided up into small channels.	The Mackenzie River has a huge delta where it enters the Beaufort Sea.
Dense	Many people living in a small area.	The population of the Toronto-Montreal area is dense.
Distillation	A process that involves evaporation followed by condensation.	Distillation is an important part of the oil-refining process.
Downbound	Toward the mouth (lowest end) of a river.	Most downbound ships on the St. Lawrence River call in at Montreal before crossing the Atlantic Ocean.
Drainage Basin	The land drained by one river and its tributaries.	The drainage basin of the Mackenzie River is extremely large.
Dredging	The removal of mud and other debris from beneath a river or lake.	Dredging is carried out in harbours to keep them deep enough for large ships.
Drifts	Horizontal tunnels in an underground mine.	Miners travel along the drifts to reach the part of the mine where the ore is being excavated.
Drought	A long period of time in which no rain falls.	Drought is often a major problem for Prairie farmers.
Economic Base	The activities of an area that bring money into a settlement. Examples include stores and factories.	Toronto has a varied economic base.
Emigrate	To leave a country in order to live in another.	Western Europe was the source of a great many people who emigrated to Canada.
Endangered Species	Types of animals, birds, etc. that are close to extinction.	The whooping crane, which spends the summer in Northern Canada, is an endangered species.
Equalization Payment	Payments paid by the federal government to help poor areas of the country.	Equalization payments are made to the Maritime Provinces.
Erosion	The wearing away of rocks or soil.	Water erodes millions of tonnes of rock in Canada every year.
Eutrophication	The process that uses up oxygen from rivers and lakes.	Eutrophication, caused by pollutants from houses, farms, and industry, is affecting many Canadian lakes.
Evaporate	Liquid water evaporates to become water vapour.	On a hot summer day water in puddles on a street evaporates quickly.
Excavate	To dig and remove material.	Rocks are excavated to create a mine.
Exhaust	To use up.	A valuable ore deposit is exhausted when it has all been mined.
Exploit	To make use of.	We exploit our timber resources to make pulp and paper.
Extensive Farming	Where a small amount of labour is used on a relatively large farm.	Large machines, such as combines, allow extensive farming for wheat on the Prairies.

Term	Definition	Example
Extinct	No longer in existence.	Passenger pigeons in Canada once numbered in the millions, but are now extinct.
Extract	To remove.	Iron is extracted from iron ore.
Fabricate	To make.	Furniture is fabricated from wood.
Farmstead	The buildings, such as the house and barn, on a farm.	Barnyard / Barn / Sheds / House
Federal Government	The central Canadian government.	The federal government is centred in Ottawa.
Feedlot	A farm where cattle are taken to be fattened before slaughter.	Feedlots are common around major Canadian cities such as Toronto and Montreal.
Flood-Risk Maps	Maps that show the extent of damage in the case of flooding.	Flood-risk maps help people to plan protective measures.
Floor Price	The lowest price for which a commodity may be sold.	Floor prices are important in the mining industry.
Flume	A trough used to carry irrigation water.	Flumes are found throughout the Okanagan Valley.
Fossil	The hardened remains of animals or plants found in rock.	Fossils by the millions can be found in some sedimentary rock.
Four Corners	See HAMLET.	
Franchise	A licence to operate or manage a shop or food outlet.	A person may have to pay many thousands of dollars to buy a franchise.
Francophone	A person whose mother tongue is French.	Most people who live in Quebec are Francophones.
Free Trade	Trade with no tariffs.	Free trade would make imported goods less expensive.
Ghost Town	A village or town that has been abandoned.	Barkerville, British Columbia, became a ghost town after the Cariboo gold rush ended.
Gill Netting	A fishing method in which fish are caught in the net close to the surface of the water.	Gill netting can be used to catch Pacific Salmon.
Glacier	A thick mass of ice that covers the land.	Glaciers can still be found in the Rocky Mountains.
GNP (Gross National Product) per Person	The amount of wealth produced in a country on average per person each year.	The GNP per person is around $230 in Burundi, Africa.
Gneiss	An example of metamorphic rock.	Gneiss often appears as wavy strips of grey, white, and pink.
Granite	A type of igneous rock.	Granite has been used in the construction of numerous public buildings in Canada.

Grid Pattern	An interlocking pattern, normally with lines crossing at right angles.	Grid patterns are used to locate places on maps.
Groundfish	A fish that feeds and lives near the bottom of the sea or lake.	Cod is a valuable groundfish around Canada's coasts.
Groundwater	The water that seeps through rocks and soil.	Groundwater is the source of water in wells.
Habitat	The surroundings that an animal lives in.	Human activity in Canada's North has altered the habitat of many wild animals.
Hamlet	Small centre containing a small number of buildings and a few basic services, such as a store.	Maskawata, Saskatchewan.
Heavy Industry	Industry that involves the use of large machinery.	Steel manufacturing is a heavy industry.
Herbicide	Chemicals used to kill weeds on farms, etc.	A farmer who uses herbicides is saved many hours of pulling weeds.
Hibernate	The sleeping of certain animals, such as bears during the winter.	Bears must eat a great deal of food before hibernating for the winter.
Hinterland	The land "behind" a port, used to ship its goods and produce to markets.	
Humus	Pieces of decaying plants and dead animals in the soil.	Humus acts as a fertilizer for growing plants.
Hydro-Electricity	Electricity produced from falling water.	Hydro-electricity produces little, if any, pollution.
Hydrologic Cycle	The continuous processes that evaporate sea water, transport clouds, cause rain, and produce rivers.	Clouds are an important part of the hydrologic cycle.
Igneous Rock	Magma when it has cooled and hardened.	Igneous rocks are used to make tombstones, which last for hundreds of years.
Immigrants	People who come to a new country to live.	In 1913 over 400 000 immigrants arrived in Canada.
Import	Goods which are brought in from other countries.	We import many tropical fruits.
Industrial Land Use	A land use where factories and manufacturing occur.	Industrial land use usually locates near a major transportation route.

Industrial Park	A planned development for manufacturing plants and factories.	Industrial parks are often attractively planned.
Industry	Work that produces a valuable product or service.	Industry provides Canadians with employment and money.
Inland Sea	A sea from which little water escapes.	Inland seas covering large parts of Canada millions of years ago became full of the valuable minerals that we mine today, such as salt and potash.
Inshore Fishing	Fishing in small boats near to the coast.	Inshore fishing employs many people along the east coast of Canada.
Institutional	Land uses such as churches, schools, and libraries.	Institutional land uses are usually located in or near residential land uses.
Intensive Farming	A great deal of labour required by a relatively small farm.	Fruit farms in the Annapolis Valley of Nova Scotia are classified as intensive farming.
International Trade	The buying and selling of goods from one country to another.	International trade is vital to Canada's economy.
Inventory	A counting or stock taking.	All industries carry out inventories to determine the value of their possessions.
Irrigated	The watering of crops or plants by people.	Irrigation allows crops to be grown in some desert areas of interior British Columbia.
Key (Legend)	The portion of a map that contains symbols and their meaning.	KEY — Railway, River, House, Barn
Killing Frost	A temperature below 0°C during the growing season that severely damages a crop.	A killing frost will turn tomato plants black.
Landed Immigrant	A person who has been accepted by the government to apply to become a Canadian citizen.	People coming to Canada are usually landed immigrants before becoming Canadian citizens.
Large-Scale Map	A map that shows only a relatively small area, but tends to show a large amount of detail.	A street map of a city is a commonly used large-scale map.
Latitude	The distance of any point north or south of the equator. It is measured in degrees.	Regina, Saskatchewan, has a latitude of 50° 30' North.
Light Industry	An industry that uses small machinery.	Making clocks is a light industry.
Line of Latitude	A line that circles the world at a constant distance from the equator.	A line of latitude passing through Edmonton, Alberta also passes through York, England.
Line of Longitude	A line that circles the world and passes through the North and South Poles.	Lines of longitude are used to set up time zones in the world.
Line Scale	A line drawn on a map to represent distances on the ground.	0 10 20 30 40 50 60 km
Longitude	The distance of any point east or west of the Prime Meridian.	Canada's longitude is west of the Prime Meridian.
Magma	Melted or molten rock.	A volcano spews out magma during an eruption.

Main Stream	The most important channel of a river into which the tributaries flow.	All river basins have the same name as the main stream.
Manufactured Goods	Products made from raw materials.	A television set is a manufactured good that almost all Canadians have.
Marginal Farmland	Land with soil that can produce only fair to poor crops.	Rural poverty is common in areas of marginal farmland.
Maximum Temperature	The highest temperature for a certain location.	The maximum average temperature for Montreal in July is 26.3°C.
Megajoule	A unit of electrical measurement.	A 100 W lightbulb burning for 10 000 s uses 1 MJ of electricity.
Megalopolis	When several cities grow to the point where they touch each other, the area is a megalopolis.	Southern Ontario is the first region of Canada where a megalopolis has begun to form.
Metamorphic Rock	Igneous or sedimentary rock that has been heated or exposed to stresses can become metamorphic rock.	Many of our minerals, such as iron, silver, and gold, are found in metamorphic rocks.
Migration	The seasonal movement of animals, birds, etc.	Caribou migrate over vast areas of Canada's tundra in search of food.
Military Grid	A pattern of intersecting straight lines used for locating places on a map.	A grid diagram with horizontal lines labelled 22 and 21 on the right, and vertical lines labelled 50, 51, 52, 53 along the bottom.
Mineral Matter	The grains of sand, silt, or clay in the soil.	Growing plants send their roots down into the mineral matter of the soil.
Mineral Rights	The right to mine in a particular area.	Mineral rights must be obtained before mining can start.
Minimum Temperature	The lowest temperature for a certain location.	The minimum average temperature for Vancouver, B.C. in January is −0.4°C.
Mother Tongue	A person's native language.	English is the mother tongue of most Canadians.
Mouth	The place where a river enters a lake or the sea.	The mouth of the St. Lawrence River contains Anticosti Island.
Multicultural Society	A society in which there are a large number of immigrant groups that retain many customs from their homelands.	In a city like Toronto the many languages spoken reflect its multicultural society.
Natural Resources	A naturally formed object of which we make use.	Forests, minerals, and fish are our most valuable natural resources.
Natural Vegetation	The mixture of plants that grows without human interference.	Most natural vegetation in southern Canada has been cleared due to human activity.
Net Exporter	A country that exports more than it imports.	Canada is a net exporter of mine products.
Net Importer	A country that imports more than it exports.	Canada is a net importer of manufactured goods.
Non-Renewable Resources	A resource that, once used, cannot be used again.	Coal is a non-renewable resource.
Nuclear Reaction	A reaction that releases radioactivity.	Nuclear reactions are used to create electricity.

Offshore Fishing	Fishing in large ships a long way from land.	Offshore fishing catches more fish than inshore fishing.
Oil Sands	A thick oily sand.	The oil sands will probably be of great value to Canada in the future.
Open-Pit Mining	Mining in a big hole in the ground.	Open-pit mining occurs when the valuable deposit is near the surface.
Ore	A rock that is worth mining.	Iron ore is mined in Labrador.
Orient	To lay out a map correctly according to direction.	A properly oriented map has its north direction pointing towards the north.
Orographic Precipitation	Precipitation caused by a movement of air over a rise of land such as a hill or mountain.	Orographic precipitation occurs very often along the coast of British Columbia.

Otter Trawling or Dragnet Fishing	A fishing method in which a net bag is dragged behind a boat, at a chosen depth.	Otter trawling is often used to catch cod.
Overburden	The useless material, such as soil and rock, that has to be removed before open-pit mining starts.	The removal of overburden is an expensive process.
Pellets	Small spheres of concentrated ore.	Pellets are made near to the mine, in a pelletizing plant.
Pemmican	Dried meat pounded to a powder and mixed with pounded choke cherries by the Plains Indians.	Pemmican was an ideal food, since it was light and easy to transport.
Per Capita Gross National Product (GNP)	The amount of wealth produced in a country by one person in one year.	Canada has one of the highest per capita Gross National Products in the world.
Pesticide	Chemical used to kill animal pests that harm crops, gardens, trees, etc.	Pesticides are sprayed on tobacco crops throughout the year to control insects that can damage them.
Petroleum	A liquid fuel extracted from the ground.	Petroleum is a very valuable mineral.
Plankton	Tiny plants and animals that live in water.	Fish are attracted to the Grand Banks because of the plankton.
Podzolic Soils	Soil that has a pale layer a few centimetres below the surface.	

Podzolic soils, which form beneath coniferous trees, are very low in nutrients.

Population	The number of people in a country.	Canada's population is increasing.
Population Density	The number of people in every square kilometre.	Population density in northern Canada is much lower than in southern Canada.
Pores	The spaces in rock or soil.	Pores in rock may be filled with water, air, oil, or gas.
Power Grid	A network of electrical lines that transports power across Canada.	The power grid is vital to our way of life in Canada.
Precipitation	Water that falls to the earth in forms such as rain and snow.	Too much precipitation in an area at one time may cause flooding.
Prescribed Burning	Carefully controlled burning of stumps and brushwood.	Prescribed burning is used to prepare forest land for replanting.
Primary Industry	Industry that removes raw materials from the land or water. Also agriculture that uses the land directly.	Primary industries provide many of the raw materials that we export.
Prime Meridian	A line of longitude that passes through Greenwich, England and is 0° of longitude.	The Prime Meridian passes through Africa as well as through the North and South Poles.
Productivity	The amount that one worker produces.	The productivity of the average Canadian farmer has increased greatly in the last 50 years.
Profile	The shape of the land as seen from the side.	A profile of the land clearly shows the high and low areas.
Profit	The money that remains after all costs are paid.	All industries aim to make a profit for the owners.
Profit Motive	The desire to make money.	The driving force for almost all work is the profit motive.
Property Taxes	Taxes collected from landowners in a certain area.	Property taxes are important because they pay for many services, such as schools, roads, and municipal water supplies.
Prorationing	The assignment of the amount of product to be produced by a mine.	Prorationing helps to keep selling prices high.
Purse Seining	A fishing method in which the net is drawn together like a bag.	Purse seining is a method commonly used by inshore fishermen.
Québécois	A French-speaking Quebecker.	Most Québécois are proud of their French heritage.
Quota	The maximum amount of a commodity to be produced.	Quotas help to keep prices high by limiting the supply of a commodity. Tobacco is an example of such a commodity.
Radioactivity	The dangerous emission of rays from a nuclear reaction.	Many people are afraid of an escape of radioactivity from nuclear reactors.
Rang	A row of settled land in Quebec's long-lot system.	Roture / Rang / Road / St. Lawrence River
Range of Temperature	The difference between the highest and lowest temperature for a certain location.	The annual range of temperature for Toronto, Ontario is 20.7° − (−6.3°) = 27°C.

Ratio Scale	A scale shown in a ratio such as 1:50 000.	A ratio scale of 1:100 000 means that 1 cm on the map represents 100 000 cm on the ground.
Raw Materials	The ingredients needed to make finished products.	The raw materials needed for the furniture industry include wood, cloth, and metals.
Redevelopment	The tearing down of old buildings, and their replacement with new structures.	Redevelopment in a city often involves the tearing down of older or slum areas.
Refining	The process that makes a substance more valuable.	Iron is refined into steel; crude oil is refined into many products.
Region	An area that has some features that are the same.	The Prairie region is much flatter and drier than the Atlantic region.
Renewable Resources	A resource that can restore itself.	Falling water is a valuable renewable resource in the production of electricity.
Reserve	Land assigned to Indians by the government for their exclusive use.	Many Indians have left the reserves in search of a new life.
Residential Land Use	A land use where people live.	Residential land uses include apartments, single-family houses, and rooming houses.
Restock	To replenish stocks that are getting low.	Many of our lakes have been restocked with fish.
Roture	An individual farm in a rang (row) of the Quebec long-lot system.	Roture / Rang Road / St. Lawrence River
Royalties	A tax paid by a mining company to the government.	Royalties from the petroleum industry are important to Alberta's economy.
Run-off (Surface Water)	Run-off, also called surface water, is the water which flows over the earth's surface.	Lake Superior is a good example of surface water.
Rural-Urban Fringe	The land surrounding a town or city that has both rural and urban land uses.	Apartments / Abandoned Farm / Lumber Yard / Housing Subdivision / Slaughter house / Drive-in Theatre
Sandstone	An example of sedimentary rock.	Sandstone is sometimes quite soft and can crumble easily.
Scale	The distance on a map representing a certain distance in the world.	Scale is the key to measuring real distances on a map.
Secondary Industry	Manufacturing industry.	Secondary industry takes raw materials and processes them further.
Section	A basic unit of measurement of an area in the Prairies containing 256 hectares.	Successful wheat farms on the Prairies must have several sections of land.
Sedimentary Rock	Rock formed in layers.	Niagara Falls flows over a cliff formed of sedimentary rock.
Seigneur	The person who had control over a seigneury in early Quebec.	Seigneurs in early Quebec were responsible for managing the settlement of their seigneury.

Seigneury	A unit of land in early Quebec that was subdivided for settlement.	Today you can still see the boundaries of old seigneuries in Quebec.
Seismograph	An instrument that records vibrations in the earth.	Seismographs are used to study rock formations.
Selective Cutting	The cutting of specially chosen trees.	Selective cutting is used to get timber for fine oak furniture.
Separatism	The philosophy founded on the belief that Quebec should become a separate country.	Many Canadians are worried that separatism would damage Canada.
Shaft	The vertical tunnel in a mine.	An elevator brings miners and minerals up the shaft to the surface.
Shaft Mining	Mining beneath the surface.	Deep deposits are extracted by shaft mining.
Side Road	A road that runs at right angles to a concession line in an Ontario township.	See CONCESSION.
Site	The local landscape, bodies of water, and surrounding land uses near where a city is set up.	Winnipeg has a very flat site on which two major rivers — the Assiniboine and Red — meet.
Situation	The location of a settlement relative to other cities, provinces, countries, and transportation routes.	An important part of Windsor's situation is its closeness to Detroit, Michigan.
Skidders	Special tractors designed to drag logs.	Skidders are used to get logs to the nearest road or river.
Small-Scale Map	A map showing a relatively large area, but only a small amount of detail.	A map of the world is one example of a small scale map.
Smelting	The first stage in processing a mineral.	Iron ore is smelted to form liquid iron.
Soil Profile	The appearance of soil layers viewed from the surface down.	
Sound	A deep sea-water inlet of the ocean.	Sounds along the Newfoundland coast are known for their beauty.
Source	The place where a stream or river starts.	The sources of many of our western rivers are in the Western Mountains.
Sparse	An area where few people live.	The population of northern Canada is very sparse.
Species	Types of animals, plants, etc.	Canada's west coast region has many species of plants that cannot be grown elsewhere in Canada.
Stockpile	A huge reserve of material.	Iron ore and coal are stockpiled at the steel mills before the Great Lakes freeze.
Stope	A hole in an underground mine where excavation is taking place.	Drilling and blasting are done in the stopes of a mine.
Strain	The particular variety of a crop such as wheat.	New strains of wheat allow Prairie farmers to set up wheat farms farther north than was ever possible before.

SOIL PROFILE
Topsoil
Subsoil
Parent Material

326

Strip Mining	Another word meaning open-pit or surface mining.	Strip mining is used in Alberta to extract shallow coal deposits.
Surface Water	Water which stays or flows over the surface of the earth.	Niagara Falls is a spectacular example of surface water.
Surveying	The accurate measurement and recording of the nature of the land surface.	Surveying must be completed before new housing can be built in major Canadian cities.
Symbol	A shape or pattern on a map that represents an object.	is a common symbol for marsh.
Tertiary Industry	A service industry.	Restaurants and shops are tertiary industries.
Thematic Maps	A map that shows detail about one topic, such as the soils of an area.	Thematic maps for Canada show information in a different form than charts and lists of statistics.
Thermal Electricity	Electricity produced by burning coal, oil, or gas.	The production of thermal electricity causes much air pollution.
370-km (or 200 mile) Fishing Zone	The area of ocean within 370 km of the coast over which Canada has control.	Foreign fishing vessels cannot fish in our 370-km fishing zone without permission from the Canadian government.
Tides	The rising and falling of the ocean.	The greatest tides in the world are in the Bay of Fundy.
Tile	A pipe with holes in it that is buried underground to drain water from the soil.	Tiles are used in draining water from fields that are too moist.
Toll	The charge that has to be paid for travelling through a canal, across a bridge, or along a road.	Tolls are charged for ships to travel through the Welland Canal.
Topographic Maps	Maps which show primarily the shape of the land, as well as features made by human and natural activity.	Topographic maps are available for most areas of southern Canada. Perhaps your house is shown on such a map.
Town	A settlement larger than a village, but smaller than a city.	Major cities in Canada were once towns.
Trade	The buying and selling of goods.	Trading occurs in all parts of the world.
Trade Balance	The difference in value between imports and exports.	In 1984, Canada's trade balance was $20 400 000 000.
Trade Deficit	When the value of imports is greater than the value of exports.	When we have a trade deficit, Canada has to borrow money from other countries.
Trade Surplus	When the value of exports is greater than the value of imports.	When we have a trade surplus, we are able to spend more on services like hospitals and education.
Trade Tariff	A tax put on imported goods.	Trade tariffs encourage people to buy Canadian made products.
Trading Partners	The countries that sell products and/or raw materials to one another on a regular basis.	Canada's most important trading partner is the U.S.A.
Transhumance	The seasonal movement of people and animals up and down a mountain.	Transhumance is practised in the interior of British Columbia.
Transhipment	The loading and unloading of commodities.	Halifax is a major transhipment point on Canada's east coast.

Tree Line	The line that separates the forests from tundra vegetation, which has no trees.	The tree line almost reaches the Arctic Ocean near the mouth of the Mackenzie River.
Tributary Streams	Streams or rivers that flow into a larger stream or river.	The Ottawa River is a tributary of the St. Lawrence River.
Trolling	Fishing using poles and lines with baited hooks while the boat moves slowly through the water.	Trolling is a method commonly used by anglers.
Tundra Vegetation	Small, quick-growing plants that grow in the Arctic.	Mosses and lichens are two examples of tundra vegetation.
Underground Mining	Mining beneath the surface, using a shaft and tunnels.	Underground mining causes less destruction to the environment than surface mining.
Universal Solvent	A liquid that dissolves many substances.	Water is valuable because it is a universal solvent.
Upbound	Up-river, toward the source.	Ships go upbound from Montreal to Toronto.
Urban Planning	A process designed to make a settlement attractive, well organized, and a desirable place to live.	Urban planning must be practised all the time to ensure the most desirable results.
Village	A settlement larger than a hamlet that could contain about 200 people.	A village contains facilities like a post office, a variety store, a church, a gas station, and a school.
Water Deficits	Insufficient water supply for greatest plant growth.	Water deficits are corrected by irrigation in the Okanagan Valley.
Watershed	The (high) land that divides one river and its tributaries from the next.	The Western Mountains form an important watershed between rivers that flow to the Pacific and those that flow to the east.
Water Surplus	Too much water.	Water surpluses often lead to flooding.
Water Table	The top of the water in the rock pores in the ground.	The water table gets lower in a dry summer.
Water Vapour	Water in its gaseous form.	When there is a great deal of water vapour in the air, the humidity is said to be high.
Weather	The day to day changes in the conditions of the atmosphere. It includes temperature and precipitation.	The weather in Sydney, Nova Scotia, on October 1, 1986 was cool and cloudy but it was not raining.
Weathering	The breaking up and rotting of solid rocks.	Weathering of rock occurs quickly in Canada during the spring.
Winch	A wheel-like device for pulling heavy objects.	A winch is used to pull fishing nets onto the boat.
Written Scale	A scale shown by a written phrase.	One written scale is: 1 cm represents 1 km.
Zoning Law	A law to control the location and type of land use in an area.	Effective zoning laws would prevent an apartment being built next to a noisy factory.

INDEX

Francophone 309
Frederiction 140
Free trade 11
French-English differences in Canada 308-309

G

Geophone 215
Ghost town 123
Gill netting 270
GNP/capita 25-26
Glaciers 37
Gneiss 33
Grand Banks 265
Granite 32
Grasslands 68
Great Lakes-St. Lawrence Lowlands 3, 35
Great Lakes-St. Lawrence Seaway 282-284
Greenwich, England 18
Grid pattern of roads 235
Groundfish 270
Groundwater 188
Group work in class 179-180
Gulf Stream 59, 266

H

Habitat of wildlife 73
Hamlet 120
Hardwood 253
Heavy industry 285
Herbicide 155
Herman Island 48
Hibernation of animals 74
Hinterland 127
Holland Marsh 167-171
Humus 69-71
Hydro-electricity 221-223
Hydrologic cycle 184

I

Igneous rock 32
Immigration 100-110
  early immigration to Canada 100-101
  reasons for immigration to Canada 100-101
  case study: The Moon family 102-103
  variation in numbers of immigrants to Canada 106
  destinations of immigrants 107-108
  occupations 109
Import 8
Improved pasture 162
Indian People 85, 95-100
  living conditions 97-100
  status Indians 100
  non-status Indians 100
  Metis 100
  modern business ventures 99
Industrial land use 137
Industrial milk 160
Industrial park 277
Industry 277-300
  primary, secondary, tertiary 278
  factors affecting location 280
  aluminum industry 284-285
  steel industry 285-291

automobile industry 291-294
  small business 295-296
  restaurants 296-300
Inland Sea 248
Innuitians 35
Inshore fishing 267
Institutional land use 136
Intensive farming 154
Interior Plains 3, 35
International Date Line 19
International trade 9-13
Inuit 85-94
  historic life 85-88
  modern life, problems 87-89
  Resolute experiment 89-94
Inventory 252-253
Irrigation 69, 166

J

James Bay project 199-201

K

Kamloops 141
Key 44
Killing frost 163

L

Labrador current 59, 266
La Grande River project 199-201
Landed immigrant 110
Landscape 31
Latitude
  definition 17
  effect on climate 55-56
  line of latitude 17-20
Leeward 58
Legend 44
Library use 179
  card catalogue 179
  call number 179
  reference section 179
Light industry 285
Line graph 10
Line scale 39-41
Lobster 269-270
Location of industries: factors affecting 280-281
Long lot system 114-116
Longitude
  definition 17
  lines of longitude 17-20
Longitude and latitude 17-20
  lines of latitude 17-20
  latitude definition 17
  lines of longitude 18
  longitude 18

M

Magma 32
Manufactured goods 11
Maps 39-48
  large scale maps 41
  small scale maps 41

regional maps of Canada 47
Marginal farmland 173
Market forces 172
Market gardening 167
Megajoules 220
Megalopolis 133
Metamorphic rock 33
Migration of wildlife 74-77
Military grid 44-45
Mineral matter in the soil 69
Mineral rights 244
Mining 241-251
  distribution in Canada 242
  exploration 243-244
  case study: Potash 246-249
  problems 250-251
Missiles 25
Mixed farming 172-175
Mother tongue 308
Mount Logan 58-59
Mountain vegetation 68
Multicultural society 101

N

National Parks 78-81
NATO 23-24
Natural resources 239
Natural vegetation 64-69
  definition 64
  regions of Canada 67
NAWAPA 196-201
Net importer 247
Non-renewable energy 210-219
Non-renewable resources 239
NORAD 23-25
North Pole 18
Nuclear electricity 223
Nuclear reaction 223

O

Ocean currents 59
  effect on climate 59
Offshore drilling for oil 214-215
Offshore fishing 267
Oil 211-219
Ontario survey system 116-117
Open-pit mining 244
Orchard farming 165-166
  Okanagan Valley 166
Ore 244
Orient a map 42
Otter trawling 270
Overburden 244
OXFAM 27

P

Pan American Wheelchair Games 27
Pattern of physical features in Canada 310
Pellet 246
Pemmican 96
Permafrost 215
Pesticide 155
Petroleum 211-219
PFRA 178

Credits and Sources

All photographs not specifically credited to another source are courtesy of the authors.

Abitibi-Price Inc.: *Figures 2-64, 8-3, 8-30, 8-31, 8-32, 8-33, 8-34, 8-42, 8-45, 9-54*
Adecon Energy Systems Inc. and Indal Technologies Inc.: *Figure 7-29*
Alcan Smelters & Chemicals Limited: *Figure 9-10*
Barringer Magenta Limited: *Figure 8-9*
Calgary Chamber of Commerce/Hallmark Photographic Services/ James R. Hall Photographer: *Figures 4-23 (a), 4-23 (b).*
Canadian Geographic: *Figure 3-9*
Canadian Pacific Transport: *Figures 1-20 (a), (b), (c), 4-20 (a), 4-29, 7-33, 8-35*
Canadian Broadcasting Corporation: *Figure 10-7*
Canadian International Development Agency: *Figure 1-33*
Canadian Pulp & Paper Association: *Figure 8-44*
Canapress Photo Services: *Figure 1-41*
Casson, Albert.: *Figure 2-73*
Cominco Ltd.: *Figure 8-24*
Curriculum Branch, Ministry of Education, Ontario: *Figure 3-20 (a)*
Deer, Michael: *Figure 3-8 (a)*
Department of Culture and Recreation, Province of Saskatchewan: *Figure 2-78*
Department of Fisheries and Oceans, Ottawa: *Figures 8-55, 8-56, 8-60*
Department of Secretary of State: *Figure 3-33*
Dofasco Inc.: *Figures 1-19, 8-15, 9-12, 9-13, 9-14, 9-16*
Donaghey, Samuel: *Figure 7-35*
Energy, Mines and Resources Canada: *Figures 2-22 (a), 2-24, 2-31, 3-10, 4-3, 4-6, 4-22 (a), (b), 4-34, 5-24, 7-34*
Falconbridge Ltd.: *Figures 8-10, 8-13*
Ford Motor Company Ltd. of Canada: *Figure 9-21*
Government of B.C.: *Figures 3-27, 4-13, 4-20 (b)*
Hannell, F.G.: *Figures 2-14, 3-6 (a), (b)*
Haskett, R.C.: *Figure 1-4 (c)*
Hawaiian Visitors' Bureau: *Figure 2-3*
Imperial Oil Ltd.: *Figures 1-17, 7-8 (b)*
Inco Ltd., Manitoba Division: *Figures 8-4, 8-25*
Interprovincial Pipe Line Limited: *Figure 7-11*

Jackson, Russ: *Figure 4-17 (d)*
Kovalik, Tibor: *Figure 3-27 (b)*
Lyle, Bruce: *Figure 6-13*
MacMillan Bloedel Ltd.: *Figures 8-37, 8-38*
McDonald's Restaurants of Canada Ltd.: *Figures 9-24, 9-25*
Michelin Tires (Canada) Ltd.: *Figure 10-4*
Mining Association of Canada: *Figures 8-6, 8-14*
National Defence Headquarters: *Figures 1-34, 1-38*
Nelson Aggregate Co.: *Figure 8-11*
Nova Scotia Power Corporation: *Figures 7-30, 7-31 (b)*
Ontario Ministry of Agriculture and Food: *Figures 4-5, 5-5, 5-13, 5-15, 5-28, 6-11*
Ontario Ministry of the Environment: *Figure 8-67*
Ontario Ministry of Natural Resources: *Figures 2-15, 6-3 (a), (c), 8-40, 8-41*
Ontario Ministry of Northern Affairs and Mines: *Figure 8-8*
Oxford University Press Canada: *Figure 2-22 (b)*
Parks Canada, Fundy National Park: *Figure 2-81 (e)*
Parks Canada, Georgian Bay Island National Park: *Figure 2-81 (d)*
Parks Canada, Jasper National Park: *Figure 1-4 (e), 1-8, 2-1, 2-17, 2-81 (g)*
Parks Canada, Kluane National Park: *Figure 2-81 (b)*
Parks Canada, St. Lawrence Islands National Park: *Figures 2-81 (c), 6-3 (c)*
Parks Canada, Glacier National Park: *Figure 2-16*
Parks Canada, Revelstoke National Park: *Figures 2-16, 2-81 (f)*
Parks Canada, Cape Breton Highlands National Park: *Figures 1-4 (f), 6-3 (b)*
Placer Development Ltd., Vancouver, B.C.: *Figure 8-16*
Potash Corporation of Saskatchewan: *Figure 8-21*
Reid, Carl, Centennial Senior Public School: *Figure 3-1 (b)*
Saskatchewan Museum of Natural History: *Figures 3-13, 3-14, 3-15, 3-16*
Stelco Inc.: *Figure 9-15*
Syncrude: *Figure 7-15 (b)*
Tourism New Brunswick: *Figures 4-12, 4-33 (a), (b)*
Travel Alberta: *Figures 4-11, 5-8, 5-10, 5-11, 7-29*
Wolfe, Thomas: *Figure 6-7*